# The War Between Trees and Grasses

*Howard Thomas*

Book title: The War Between Trees and Grasses
Author: Howard Thomas, www.sidthomas.net/wp

Published by Howard Thomas, www.plantsenescence.org
Printer: Cambrian Printers, Aberystwyth
Publication date: 2017 (first edition)

*For the Abominations, and Jonathan*

# CONTENTS

Author's note: *My* is the abbreviation for million years, and *Mya* for million years ago. Similarly *Ky* and *Kya* for thousand years. Appendix 1 is a geological timeline, and other specialised terms and abbreviations are defined in the index. Many trees and grasses were harmed, and harmed each other, during the making of this book.

# PREFACE

It's a population of trees
Colonising the old
Haunts of men; I prefer,
Listening to their talk,
The bare language of grass
To what the woods say,
Standing in black crowds
Under the stars at night
Or in the sun's way.
The grass feeds the sheep;
The sheep give the wool
For warm clothing, but these–?
I see the cheap times
Against which they grow:
Thin houses for dupes,
Pages of pale trash,
A world that has gone sour
With spruce. Cut them down,
They won't take the weight
Of any of the strong bodies
For which the wind sighs.

RS Thomas *Afforestation* (1963)

RS Thomas, the very personification of the windswept, ascetic Celtic prophet, wrote constantly about grass – 'unchristened grass' he called it once – and about the 'tree of science', the 'tree of poetry', the 'tree of man' and the 'tree of Good and Evil'. The deep, and deeply felt, ambiguities of the poet's reflections are rooted in the history of a three-fold relationship between humans, trees and grasses. A vast literature, with arguably *The Golden Bough* at its epicentre, speaks of the profound religious, mystical and ritualistic significance of trees. As for people and grasses, the story of agriculture is one of the shaping of civilisation and human evolution by the cultivation of graminaceous crops. An understanding of the third element in the ménage-à-trois, the aggressive and often violent relationship between forest and grassland, requires a journey to the earliest period of life on land.

This book begins by asking where did the first forests come from, what does it take to make a tree and how has the tree life-form become refashioned by evolution as the terrestrial environment varied over geological time. The later arrival of grasses on the evolutionary scene was both a response to, and an agency of, new and potent environmental challenges - identified by the title of this book, with only a slight degree of melodrama, as a declaration of war. Much later, evolution propelled *Homo sapiens* into the midst of the age-old conflict between grass and tree, a conflict that rages to this day and which, as the words of RS Thomas attest, has shaped the nature of human psychology, physiology and culture.

## PART 1 TREE

Going up that river was like travelling back to the earliest beginnings of the world, when vegetation rioted on the earth and the big trees were kings.

Joseph Conrad *Heart of Darkness* (1899)

- The story starts half a billion years ago, when multicellular terrestrial life had yet to appear.
- The greening of the Gondwana landmass began in the Ordovician period.
- From lowly, tentative beginnings, the immigrants developed into giants: the trees had arrived.
- Fossils of the earliest land plants, and genetic analysis of living representatives of their ancestors and successors, show that the capacity to make wood was already present in the pioneers.
- Eventually after making landfall, plants began to make wood - lots of it.
- From the earliest times, trees controlled their architecture by discarding foliage and branches on a vast scale.
- Trees owe their structure and survival to the programmed disposal of most of the cells from which they're made.
- Plants do things their own way, exemplified by the distinctive growth, development and longevity of the tree form.
- Morphology is developed by the proliferation of self-similar units that are semi-independent and in competition with each other for resources.
- Plants seem to be able to reset the biological clock without being obliged to resort to the cellular spring-clean that accompanies meiosis and fertilisation.
- The meristems of trees and long-lived clonal plants may have been producing new cells and tissues for a thousand or more years, and may be potentially immortal.
- Harmful somatic mutations are usually purged by competition with other cells in the meristem population.
- Occasionally a somatic mutation might confer some kind of selective advantage, or be of horticultural value.
- Through the primal combination of morphological and physiological traits, pteridophyte forests established themselves as the dominant global vegetation type from the mid Devonian (around 380 Mya).
- They prevailed up to and beyond the time of the appearance and diversification of angiosperms in the Cretaceous period from about 150 Mya.

## Chapter 1 Landfall

Life on Earth started under water. When did living organisms first venture onto land? The answer depends on who you ask. Some microbiologists argue that the present-day ubiquity and resilience of microbial lifeforms, together with the overlap between aquatic and non-aquatic palaeo-environments, make it likely that there were terrestrial microbial mats and crusts more than one billion years ago (Battistuzzi et al. 2004, Beraldi-Campesi 2013). Measurements of isotope ratios in Precambrian sediments provide some support for such an antediluvian origin of life out of water (Knauth and Kennedy 2009). But in the absence of convincing fossil evidence from this period, most biologists and geologists think that multicellular life capable of more than transient survival on land dates from no earlier than about half a billion years ago. The emergence of ancestors of the organisms that make up today's terrestrial ecosystems marks a decisive break with history up to that point, by initiating a fundamental transformation not just of biology but of the planet itself.

A narrative familiar from television documentaries and popular natural history publications (for example Attenborough 1979) describes mudskipper- or coelacanth-like fish venturing into habitats at the water's edge and rapidly evolving into tetrapods. Sometimes there are references to features thought to represent the fossilised burrows and tracks of invertebrates that pre-date invasion by the fish. But awareness of the evolutionary origin of terrestrial animals somewhat overshadows that of the land flora, at least in part because of a parochial preoccupation with human origins - in the words of Ned Friedman 'Zoologists talk about the importance of the Cambrian explosion (542 Mya) in terms of human evolution, but humans would not have evolved without land plants' (Plackett and Coates 2016). By making landfall before animals gained a secure foothold, plants were mining mineral nutrients, creating the first soils, altering the composition of the atmosphere and establishing the autotrophic foundations of food chains that made evolution of the terrestrial fauna possible (Bennici 2008); but most people don't know this because plant evolution is hardly taught any more; and zoology has much better PR than botany.

The Gonds are a tribe that settled in eastern Madhya Pradesh from the 13th to the 19th centuries. They gave their name to the Gondwana region of India: Gondwana is Sanskrit for 'forest of the Gonds'. In 1885, the geologist Eduard Seuss wrote of a southern supercontinent dating from more than 700 Mya 'We name it Gondwána-Land, after the common ancient flora of Gondwána'. The fossil record identifies this as the land onto which the earliest multicellular lifeforms scrambled (Wellman 2010). Between 400 Mya and 300 Mya, Gondwanaland joined with the northern Laurussian block to make the Pangaea supercontinent, which subsequently fragmented into the tectonic units that form the present continental landmasses. These large-scale geological events were accompanied by, and were to some degree the causes of, a syndrome of global environmental change that drove evolutionary processes including colonisation of the land and dispersion across terrestrial environments. A combination of tectonic movements, orogenesis, the glaciation cycle, fluctuating sea levels and variations in ocean salinity had profound effects on aquatic life and the incentive to make landfall. For example, the concentration of halides in the modern ocean is about 3.4%. It is estimated that, during the Precambrian (>550 Mya), salinity was much higher - around 5%, but falling rapidly towards the Cambrian (550-500 Mya), at the dawn of the period when the land was being colonised (Hay et al. 2006). Carbon cycle models estimate that, between 500 Mya and 450 Mya, atmospheric $CO_2$ concentration decreased from about 20 times to 8 times present-day values (Royer et al. 2004). In the same period the concentration of atmospheric $O_2$ increased from about 4% to close to the modern figure of 21%, an oxygenation event created by the earliest land plants themselves (Lenton et al. 2016). It isn't necessary to buy into the whole autoregulatory Gaia principle (Lovelock and Margulis 1974) to appreciate that plants have prodigious powers of building and maintaining their own amenable habitats: the terrestrial pioneers clearly moulded their environment on a planet-modifying scale. Through humification and the weathering of rocks to supply minerals, early plants were active participants in converting palaeo-soils into sources of nutrients and water, and into substrates for anchorage (Xue et al. 2016).

The structural and developmental features that fitted the early land plants for the challenges of their new environment established the foundation of the body plans and lifecycles of their present-day descendants (Rothwell et al. 2014). Notable among these traits are polarisation of the axis into root and shoot, rebalancing the alternation of generations towards prominence of the sporophyte, elaboration of scales and microphylls into leaves, differentiation of epidermis and stomata and installation of conduits for moving water and nutrients between different tissues. The pioneer plants, descendants of streptophyte algae (Zhong et al. 2013), may have been little more than twiggy or cushion-like in form but, by the Middle Ordovician (about 470 Ma), they were well on the way towards full adaptation to a terrestrial existence: to quote Eldred Corner (1964), they had become 'dried off on the surface, waxed against evaporation, rooted, piped for water flow (and) built up with transparent bricks made from the excess of sugars'. The timing of evolutionary events is subject to regular revision in the light of new fossil discoveries and proxy analyses, but current thinking places the first appearance of vascular plants during pre-Silurian (late Ordovician) or Early Silurian times (about 450 Mya), seed plants (including trees) in the Middle Devonian (around 390 Mya) and flowering plants 160-150 Mya, in the Late Jurassic–Early Cretaceous (Silvestro et al. 2015).

## From pioneers to forests

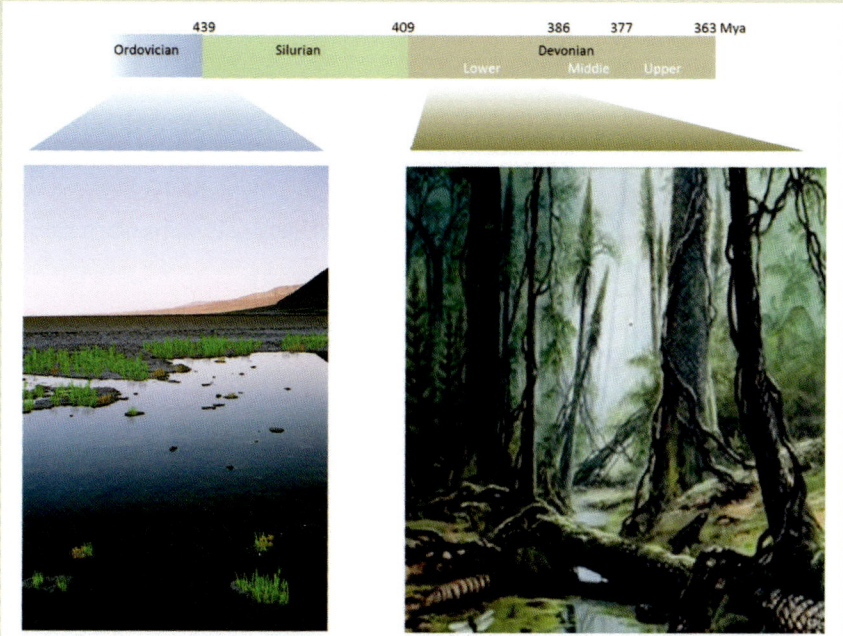

The greening of the terrestrial environment began almost half a billion years ago (Willis and McElwain 2014). Fossilised remains derived from the first land plants include spores from around 475 Mya, sheets of cuticle from 450 Mya, stomata (420 Mya), root-like structures around 408 Mya and complex vascular and supportive tissues (380 Mya). The earliest plants, such as *Cooksonia*, *Aglaophyton* and *Rhynia*, were small in stature with leafless branching stems, bearing mostly terminal, but in some cases lateral, sporangia, and were anchored by rhizoids or rhizomes. By about 385 Mya the first arborescent forms of spore-producing plants were established. It is thought that leafless tree fern-like species such as *Pseudosprochnus* (which became extinct in the late Devonian) represent the link between the earliest vascular plants and the first leaf-bearing trees. The earliest forests comprised lycopsids (giant club-mosses), sphenopsids (horsetails), filicopsids (tree ferns) and progymnosperms (for example *Archaeopteris*). The progymnosperm group has no extant representatives but is considered to be the source from which all modern seed plants evolved.

## Chapter 2 Arboreal origins

I dwell on conditions experienced by the earliest land plants, and the adaptations adopted by these pioneers, because they have a direct bearing on the evolutionary origin of the tree lifeform. Indeed, it could be argued that the appearance of the primeval forest biome was an inevitable destination once plants began to travel along the road from the aquatic environment to a fully terrestrial existence. From lowly, tentative beginnings, the immigrants proceeded to develop into giants. Let us look more closely at the structural and developmental factors that make a tree a tree.

Many adaptations combined to make the arboreal life-form. One of central significance is the strategic deployment of senescence and death (Thomas et al. 2009). Plants don't fossilise well, and the record of the avant-garde is patchy; nevertheless, such specimens as we have suggest that the earliest land plants recruited selective elimination of cytoplasm (a cell process as ancient as life on earth – Ameisen 2002) to become an essential part of their developmental and reproductive repertoire. Continued access to water is obligatory for life on land. In the first terrestrial plants, water and nutrients were moved around through the prototypes of vascular systems, consisting of hollowed-out cells arranged in pipeworks. Tracheid-like cells with characteristically thickened cell walls, secondary xylem and central pipework bundles are clearly visible in the anatomies of fossils dating back to the dawn of the terrestrial era (Edwards 1993, Edwards et al. 2006, Kenrick et al. 2012). These, and other structures in the fossil record, tell us that the capacity to make tubes by the programmed elimination of cell contents, and to thicken the walls of these empty cells, was already present in the non-vascular pioneers. We can be sure of this because the genes encoding the necessary biochemical systems are ancient, long predating colonisation of the land (Thomas et al. 2009, Martone et al. 2009, Weng and Chapple 2010, Lucas et al. 2013). Thus the full physiological toolkit to make wood was ready to be accessed quite soon after making landfall.

The early land plants found themselves in an environment rich in essential resources – light, carbon dioxide and warmth – and virtually free from predators, disease and competition. Once the problems of accessing, allocating and conserving water and nutrients had been solved, photosynthesis took off in a big way. There was an explosion in vegetation productivity, with the result that plants evolved relatively rapidly into the largest of living organisms: trees. By 370 Mya, the primeval forest was the dominant terrestrial biome. Some authorities believe that, once they were fully adapted to the hydraulic and nutritional challenges of the new terrestrial environment, plants (including their modern descendants) became incontinent photosynthesisers. In 1977 John Harper wrote that 'the green plant may indeed be a pathological overproducer of carbohydrates.' Some (me included) have even proposed that wood originated as a gigantic dump for excess carbon fixed by an over-active photosynthetic machinery that had originally evolved to work optimally in the dim, cool underwater environment (Thomas and Sadras 2001).

One effect of the large increase in atmospheric oxygen concentration associated with plant terrestrialisation was to allow early animals to grow to sizes no longer seen in the modern fauna. For example experimental studies support a direct relation between oxygen partial pressure and gigantism in Palaeozoic (>300 Mya) insects (Harrison et al. 2010). Green plants may be oxygenic, but this doesn't stop vegetative and reproductive development being constrained under anoxic conditions such as flooding (Fukao and Bailey-Serres 2004). The tree form represents a solution to the problem of oxygen limitation in bulky organs: you can grow as massive as you like if most of your cells are dead. Another way in which plants have been able to modify their morphologies to meet environmental and developmental demands is by basing their lifestyles on built-in obsolescence.

It is often stated that the early terrestrial flora was predominantly evergreen, a habit associated with nutrient-poor environments such as those experienced by the colonisers (Aerts 1995). This may be so, but contrary to some assertions, evergreen does not mean that foliage neither senesces nor abscinds (Wyka and Oleksyn 2014). Whether one takes a moderate or extreme view of the origin of the tree form, there is a strong case to be made that the interaction between the earliest vascular plants and their new environment encouraged structural redundancy and other profligate habits (Thomas and Cleal 1999, Thomas and Sadras 2001). We can see this in the extravagant throw-away behaviour of early trees, a habit that persists to this day in temperate deciduous forests (Kikuzawa and Lechowicz 2011). The earliest known forest tree, the Gilboa tree (*Pseudosprochnus*), from 385 Mya, pre-dated the evolution of the first leaf-like lateral appendages (Boyce 2010) and is thought to have photosynthesised through green non-fertile branches. The fossil record shows these branches to have been actively dropped (Stein et al. 2007). The purposeful shedding of twigs and branches is characteristic of maturity in many modern tree species, including larches, pines, poplars, willows, maples, walnuts, ashes and oaks. The technical term for active abscission of branches is *cladoptosis*. Looy (2013) has discussed the origin of cladoptosis in early conifers, dating from more than 300 Mya, and suggests an association between increasing size and the capacity for architectural self-modification. Interestingly there is also evidence that the production of abundant litter had an effect on the occurrence of wildfire, a frequent source of environmental disturbance at a time of high atmospheric oxygen concentrations. We will return to the subject of fire as an agent of ecological change when grasses introduce themselves, much later in the evolutionary timeline. Primitive trees such as *Archaeopteris*, which flourished around 355 Mya, controlled the architecture of their canopies by the immoderate shedding of lower megaphylls, the evolutionary precursors of true leaves (Raven 1986). Fossilised leaf litter banks show that many *Glossopteris* seed fern species, which arose about 300 Mya in the supercontinent Pangaea, were deciduous on a grand scale (Plumstead 1958, Pigg and Trivett 1994). Some authorities have introduced a note of caution (see Gulbranson et al. 2014, for example); but on the whole it seems reasonable to conclude that the selective death and elimination of cytoplasm, cells, tissues and organs played a crucial part in establishing the morphologies, sizes and life-cycles of trees from the earliest period of their evolution.

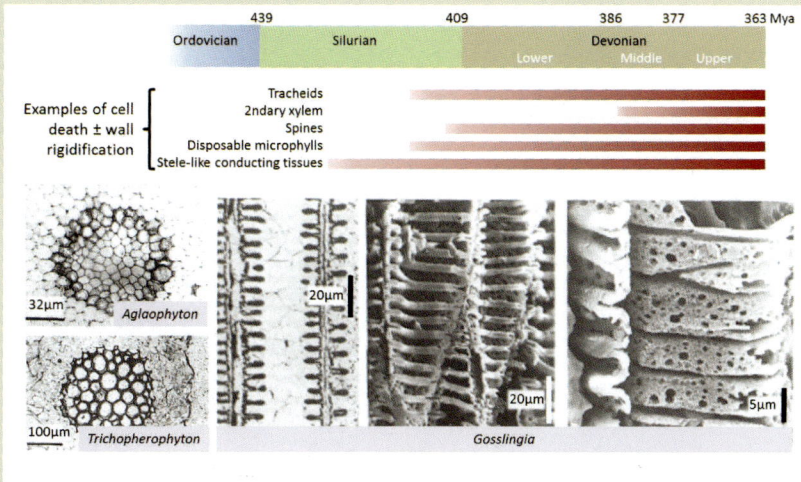

## The ancient capacity for developmental cell death and making wood

The fossilised remains of the earliest land plants show that selective death of cells was an established feature of development and of adaptation to the terrestrial environment. *Gosslingia breconensis*, a zosterophyllous species from the Lower Devonian (>400 Mya), possessed tracheids, xylem cells from which cytoplasm is eliminated by programmed cell death. The characteristic patterns of secondary thickening are preserved during coalification (Edwards 2003). *Aglaophyton major* and *Trichopherophyton teuchansii*, from the Lower Devonian Rhynie chert (380 Mya), are examples of early tracheophytes in which conducting tissues are organised into a central stele with thickened cell walls (figures from Edwards 1993). Other instances of programmed cell death in the early fossil record include microphylls (proto-leaves) that were actively shed, and spines. A common feature of the genetic regulation of selective cell death and wall thickening is the control network exerted through transcription factors of the NAC family. There is growing evidence from comparative genomics that acquisition of the capacity for NAC regulation was a critical step in the evolution of land plants from their non-vascular ancestors (Xu et al. 2014).

**The ancient habit of shedding leaves**

Fossilised leaf litter beds of early trees

Archaeopteris                    Glossopteris

There are many examples of fossilised leaf litter beds, such as those from the Devonian progymnosperm *Archaeopteris* (>360 Mya) and the seed fern *Glossopteris* (<300 Mya). *Archaeopteris* is considered to be one of the earliest modern trees (Meyer-Berthaud et al. 1999) and was the dominant form in Devonian forests. By enriching the nutrient levels in soils and the aquatic environment, its habit of abscission, actively shedding leaves and branches, was one of the factors influencing profound geochemical and evolutionary changes in the Devonian (Algeo and Scheckler 1998). Glossopterids dominated the forests of Gondwana during the Permian. They were prodigiously deciduous, leaving thick mats of fossilised leaves, and were major components of coal deposits in the Southern Hemisphere (Taylor et al. 2009, Holdgate et al. 2005). Leaf litter continues to have a global influence on nutrient and energy relations in the biosphere (Pietsch et al. 2014).

## Chapter 3 The sum of the parts

Plants in general are made from disposable parts (Thomas 2016). The limited lifespans of these parts are integral to their developmental programs. Senescence (which in the plant business means something different from ageing) is the terminal phase of development that usually (but not inevitably) leads to death of a cell (Thomas 2013). Trees in particular owe their form and survival to the programmed senescence and death of most of the cells from which they're made. All of which raises a challenging question of fundamental biological importance: what is the relationship between the lifespan of the whole organism (measured in hundreds or even thousands of years for the longest-lived trees) and that of its component parts? It's a thought-provoking point – is the endurance and lifestyle of trees in spite of or because of their indulgence in wholesale cell death? This, a matter central to the question as to what makes a tree a tree, is difficult to tackle head-on, but we can gain insights by approaching the issue from a few different directions.

We start by asking: what is an individual? Is a tree a single organism or more like a colony or population of associated but existentially independent units? A close look at tree architecture gives insights into what selfhood or singularity means for the arboreal habit. The idea that plant structure is modular can be traced back at least as far as Johann Wolfgang von Goethe. On 17 May 1787 in a letter from Naples, Goethe wrote:

> I am very close to discovering the secret of the creation and organization of plants ... The *Urpflanze* is to be the strangest creature in the world; Nature herself shall be jealous of it. After this model it will be possible to invent plants ad infinitum, which will all be consistent, that is, they could exist even if they have no actual existence ... And the same law will be applicable to everything alive.

The generative nature of Goethe's *Typus* exerted a profound influence over the development of natural philosophy in Germany and beyond, with repercussions for the course of everything from evolutionary thought to music theory (Heller 1961, Don 1996). Synthesis of the modular concept and genetics in the modern era probably began with Bateson (1894). The body plan of multicellular organisms consists of repeated segments. The segments that make up the plant shoot system are termed phytomers (Buck-Sorlin 2013). The fractal organisation of repeated self-similar phytomeric units is central to current approaches in structure-function modelling of plants (Vos et al. 2010). What are the factors that specify the form of each module and their assembly into the whole plant? Driven by his restless curiosity, Leonardo da Vinci, no less, brought an artist's eye and a scientist's mind to bear on this question. Contemplating 'the relative thickness of the branches to the trunk', he wrote in the *Notebooks*: 'All the branches of a tree at every stage of its height when put together are equal in thickness to the trunk [below them]. All the branches of a water [course] at every stage of its course, if they are of equal rapidity, are equal to the body of the main stream.' This admirable leap of imagination, likening the morphogenesis of a tree to the form of a river delta (McMahon and Kronauer 1976), has given rise to the da Vinci rule, a principle in the physical modelling of tree architecture (Minamino and Tateno 2014).

Hallé (1986) has described several variations on the theme of modularity and its implications for plant development. One concerns size: modules may remain singular and become large (as in some trees like *Araucaria* spp., and few long-lived herbaceous monocarps such as *Agave*); or be singular and miniaturised (to be discussed further when we consider the origins of the herbaceous habit); or may retain more or less the same size but increase in number, thereby developing into large individuals - the mode of architecture of most tree species. I've already touched on environmental and physiological factors driving the early evolution of gigantism in trees. Close examination of module repetition processes during tree morphogenesis reveals some extraordinary features of growth and development.

Hallé uses the term 'reiteration', first introduced by Oldeman (see Hallé et al. 1978), to describe a particular type of repetition, seen in many tropical (for example mango) or temperate (oak, beech) trees, that usually starts with the activation of a dormant adventitious bud located on the trunk (Fink 1983). The meristem becomes active, leading to the bursting bud behaving like a germinating seed (elsewhere – Thomas 1994 – I have described resting buds as 'essentially seeds that never fell to earth'), not only extending a shoot with foliage, but also sending a root downwards, inside the bark of the parent tree. After passing through a phase of hemi-parasitism, it becomes a new, rooted individual, genetically identical to, and fused with, its parent. Reiteration can be traumatic in origin, as seen in the regeneration of clonal shoots (ramets, commonly called suckers) from a wounded, pollarded or coppiced tree, or from a stump; or it can be adaptive, occurring at a time and place determined by environmental and internal physiological factors. Either way, this is not branching in the sense of outgrowth of laterals from axillary buds but true clonal growth. It's a mode of construction in which each module is functionally semi-independent and in competition with others for resources. This was recognised long ago by Immanuel Kant, writing in 1790, the same year as Goethe's *Metamorphose* was published: 'each branch or leaf of a tree may be regarded as merely engrafted or inoculated upon it, and therefore as a tree with an existence of its own, simply attached to another from which it nourishes itself' (quoted by Arber 1950). Yet the plant clearly sustains a gestalt, an integrated individuality to which the constituent modules are ultimately subordinate. Intrinsic to selfhood is proprioception, an organism's sense of its own form in relation to the environment (Hamant and Moulia 2016).

Self-perception of modular structure is a factor mediating in size, which in turn is intimately related to ageing and lifespan. Trees are big and long-lived, and of course organisms that are large and old and that stand in the same place for a long time will also be weatherbeaten, the visible signifier of ageing. We often revere ancient trees because the experience of their long lives is written in the gnarly organic contortions of their forms. Across the world there are countless examples of old trees of distinction that have seen action and require the support of crutches and buttresses. One such is Darwin's mulberry, believed to have been planted in his garden at Down House, Kent in 1609 (Armstrong 2009). It's hollow, propped up with several brackets, and at one time was filled with concrete to shore it up. And yet it still flowers and sets fruit every year. By one measure – morphology – it has aged greatly. By another – fecundity – it has hardly aged at all.

Size has consequences, too, for physiology as well as form. It is a significant factor in the development of symptoms of ageing-related deterioration (Day et al. 2002). Size affects nutrient allocation, the ratio of photosynthesising to respiring tissues and the hydraulic functions of the vascular system (Mencuccini et al. 2005). But a definitive mechanistic link between size and ageing has proven difficult to establish. Declining photosynthesis in some species is associated with increasing distances between the roots and the extremities of the crown and with decreases in the flow of water and sap in the whole tree (Hubbard et al., 1999). On the other hand Lanner and Connor (2001) found no evidence for age-related deterioration in the vascular function of bristlecone pines over the age range 23 to 4713 years. Moreover, although there is a general inverse relationship between growth rate and increasing age class in trees, growth rates in the oldest and largest trees are frequently sustained for the remainder of their lives (Johnson and Abrams 2009). The vigour of the oldest trees is reminiscent of a famous demographic paradox, that of 'negative senescence'. According to life-tables (see, for example Benjamin and Overton 1981), if you can survive beyond a threshold age for deaths attributable to biological events in the ageing syndrome, your chances of death appear to decrease (this pattern is a source of comfort to those in my age-bracket, even if it is a deception). In the words of the poet Frederick Seidel:

> At seventy-seven I reached my prime.
> But seventy-eight was absolutely great.
> And then came fab seventy-nine and continuing to climb
> I upped my wingbeats to an even higher rate.
> *Winter Day, Birdsong* (2016)

It's easy to underestimate how much control plants can exert over their own forms and futures. The morphogenetic constraints of modularity ensure that trees do not blindly expand outward into space. Crown productivity and longevity in trees may be sustained by the process of adaptive reiteration, which decreases the ratio of respiring to photosynthesising tissues, enhances hydraulic conductance to newly developing foliage, reduces nutrient loss, rejuvenates apical meristems and increases lifetime reproductive output (Ishii et al. 2007). Many species will reshape an established architecture by invoking selective cell and organ elimination which, as we have seen, is a physiological capacity as old as the colonisation of land by the first plants.

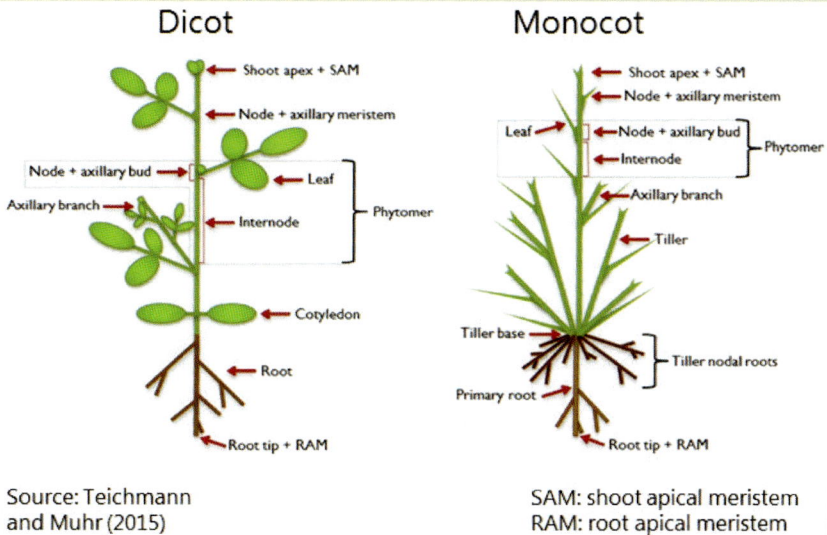

**Plants are organised on modular lines**

Dicot — Monocot

Source: Teichmann and Muhr (2015)

SAM: shoot apical meristem
RAM: root apical meristem

In their review of the modelling of plant form, Prusinkiewicz and Runions (2012) listed the objectives of computer simulations of morphology as: description of form; analysis of causality; analysis of self-organisation; decomposition of problems; hypothesis-driven experimentation; and integrative view of development. Central to the modelling approach is the concept of plant architecture as modular and recursive (Teichmann and Muhr 2015). At its most reductive, plant structure comprises shoot modules (phytomers) and root modules. In the characteristic way it occupies 3-dimensional space, a tree, being a long-lived relatively slow-growing organism, embodies its architecture and the morphogenetic pathways that led to it. Models that seek to establish a mechanistic basis for tree forms often assign a prominent role to competition between modules for space, light or internal resources (for example Palubicki et al. 2009). The evolutionary history of trees is a record of reconciliation between the individual organism and the semi-independence of its component structural modules (Hallé et al. 1978).

## Chapter 4 Lifespan

The subject of age in relation to modular and clonal behaviour, particularly in trees, is a complicated one. Reiteration, development of polyaxial architecture, repeated formation of ramets, coppice growth and so on are all variations on the theme of clonal propagation. Many plants that form clones by asexual reproduction can proliferate to establish community-sized 'individuals' of extraordinary longevity - maybe, as is the case for the Proteacean shrub *Lomatia tasmanica* (Lynch et al. 1998), in excess of 40 Ky. A clonal cluster of genetically identical aspen trees in Fishlake National Forest, Utah ('Pando' or 'The Trembling Giant') may be twice as old as *Lomatia* (DeWoody et al. 2008, Sibley 2009).

Some of the meristems of ancient individual trees will have been generating cells, tissues and organs for millennia. Even a cell replication mechanism of the highest fidelity would be expected to propagate a significant number of errors over such an extended timescale (de Witte and Stöcklin 2010). Somatic mutations can also be a consequence of genetic and metabolic damage caused by reactive oxygen species and free radicals, factors that tend to build up with age (Munné-Bosch 2007). The rate of somatic mutation has been used as a clock, to estimate lifespan. A study of a 10 Ky-old natural clonal population of *Populus tremeloides* in British Colombia reported an average of 8% (with wide error limits) decline in pollen fertility per 1000 years (Ally et al. 2010). The authors ascribe this deterioration to somatic mutation or epigenetic effects or both. On the other hand, analysis of bristlecone pine plants up to 4,700 years old found no statistically significant relationship between age of individual and frequency of mutations in pollen, seed and seedlings (Lanner and Connor 2001).

The association between reiterative architecture and extreme longevity brings us to the subject of the radical divergence between plants and animals in the way they are organised and develop. This difference corresponds to a fundamental contrast in the relationship between lifespans of parts and ageing of the whole organism. August Weismann (1893) proposed the concept of soma and germline and argued that, by implication, death in multicellular organisms must be programmed. During early embryogenesis in animals, somatic cells (destined to give rise to all body tissues other than those producing eggs and sperm) are differentiated from germ line cells (committed to becoming the gametes and reproductive organs). Normally, once a cell is committed to the germ line or soma, neither it nor the cells derived from it can divert into the alternative developmental fate. As a consequence, the animal reproductive system is not exposed to the developmental signals that subsequently induce formation of the major somatic tissues and organs, and mutations can be inherited only if they affect germ line cells. But there is no distinction between soma and germline in plants. Plants do things their own way – evident in features of development and lifecycle such as alternation of generations (Walbot and Evans 2003) and totipotency.

Most viable plant cells are potentially totipotent - that is capable of changing into another cell type even when fully differentiated. An example is the transdifferentiation of mature leaf cells into xylem cells. Several plant species can pull off this trick, but the star performer is *Zinnia* (Escamez and Tuominen 2014). Intact green mesophyll cells of *Zinnia* leaves can be physically isolated into a culture medium simply by using a blender or a pestle and mortar. Provided the right ingredients are present in the medium, the photosynthetic cells will de-differentiate over a period of a few days. Then they will go through the developmental stages leading to the differentiation of tracheids – the hollow xylem cells of wood observed in the earliest land-plants, as I described earlier. Another aspect of totipotency is the capacity to reverse the developmental clock. In some (maybe most) species, mesophyll cells in a state of advanced senescence can rewind and revert to the juvenescent state (Zavaleta-Mancera et al. 1999).

Meristems have special properties that sustain the long-term integrity of their proliferative functions. Meristem mitosis appears constantly to reboot juvenility and this makes shoot apical meristems inherently perennial or even eternal (Salguero-Gomez et al. 2013). Principles that govern the ageing behaviour of animal cells (for instance, the Hayflick limit on the number of cell doublings - Hayflick and Moorhead 1961) do not apply. And apart from a few unconvincing reports, there is no evidence from plants in favour of the related cytogenetic mechanism in which telomere attrition leads to chromosome instability and genome breakdown. The fundamentally different responses of plants and animals to telomere disruption are probably a consequence of contrasting developmental and genomic architectures. In the words of McKnight and Shippen (2004), plants have 'an amazing capacity to withstand raging genomic instability'. Progressively decreasing allocation of resources to repair and maintenance (an implication of the Weismann principle) is often considered to be intrinsic to ageing-related physiological decline – the Disposable Soma theory of ageing (Kirkwood 2002). Limited material and energy resources may be an issue for heterotrophs, but autotrophs are different. Some authorities have argued that meristem organisation, taking the form of a stratified cytostat (closely controlled continuous culture), allows the most deleterious somatic mutations to be efficiently purged by selective pressures acting on the population of newly derived cells (Klekowski 2003). Moreover, the high-UV conditions encountered by the first land plants would have strongly selected in favour of intrinsic genome resilience (Rozema et al. 2002, Popper et al. 2011). It certainly seems like plants have evolved constitutional and organisational measures to cope with ageing-related threats to genome integrity without needing to resort to the kind of p53-mediated cell death response seen in animals (McKnight and Shippen 2004). There is even evidence that genetic mosaicism within meristems may be an important source of adaptive fitness for a long-lived organism, generating tissues with new genotypes better adapted to variable environments (Folse and Roughgarden 2012). And not just adaptation to the natural environment: sometimes, like ruby grapefruit (Friend 1934) and many variegated garden plants (Friedman 2013), useful somatic mutations can be perpetuated through human selection.

## The vast ages of individual trees and clonal plants

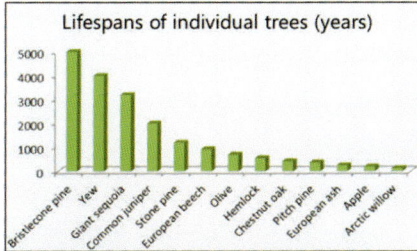

Lifespans of individual trees (years)

Ywen Llangernyw (The Llangernyw Yew)

King's Holly (*Lomatia tasmanica*)

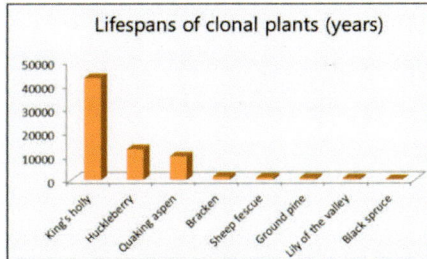

Lifespans of clonal plants (years)

Living ancient plants are of historical and cultural importance (Dujesiefken et al. 2016) and are often protected to some degree. Until such a plant dies and can be analysed by, for example, tree ring dating (Schweingruber 1988), its age must usually be estimated indirectly and non-invasively (Lennon 2009, de Witte and Stöcklin 2010). The ages of bristlecone pine trees in the Methuselah Grove located in the California White Mountains are estimated, with high confidence, to be greater than 4000 years (Lanner and Connor 2001). Ancient yews are a feature of graveyards throughout the UK (Moir et al. 2013). It has been claimed that a specimen in the North Wales village of Llangernyw might be the oldest tree in Europe and could rival the bristlecone pine in longevity, but this is disputed (Kinmonth 2006). Yews are the quintessential reiterative species (Hallé et al. 1978), clearly developing through polyaxial fusion and often becoming fragmented over their long lifespans, which makes definitive determination of age difficult. King's holly is a clonal Proteacean shrub confined in the wild to a small area of the Tasmanian Wilderness World Heritage Area. The plant is a sterile triploid and is an endangered species vulnerable to fungal disease and wildfire. Perenniality is a trait as old as the earliest terrestrial biomes and is the evolutionary source of the shorter lifespans of seed plants in the modern flora (see Corner 1964, Friedman and Rubin 2015).

**Shoot apical meristems are resilient and rarely propagate somatic mutations**

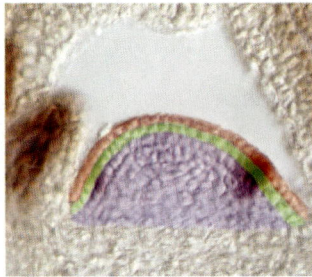

Shoot apical meristem
Modern angiosperm

Shoot apex
Early Devonian (405 Mya)
embryophyte

Ruby grapefruit
Meristem somatic mutant

This pseudocoloured micrograph of a vertical section through a typical angiosperm shoot apical meristem shows a series of distinct cell layers. Although it often is possible to assign fates to particular regions of the apex, the plasticity of plant morphogenesis means that cells can be readily induced to switch lineage in response to experimental treatments or changing physiological conditions. Fossil shoot apices show that formation of branches and leaf-like laterals is an ancient morphogenetic feature (Gerrienne et al. 2001). By integrating the results of phylogenetic and genomic studies and applying Haeckel's Biogenetic Law (Niklas et al. 2016), the evolution of meristem function in shoot development can be recreated (Sanders et al. 2007, Frank et al. 2015). The combination of structural stability, stress tolerance, retained totipotency, unlimited cell division potential and capacity to purge genetic errors makes the plant meristem effectively immortal. The occasional somatic mutation makes it through to the phenotype if it does not compromise the fitness of the whole plant. In some cases, such mutations may even improve fitness, and are sources of traits attractive or useful to horticulturists. 'Ruby' varieties of grapefruit and other citrus species have arisen as bud sports, the consequence of transposon mutagenesis events at the apex (De Felice 2009).

## Chapter 5 From trees to herbs

Between about 400 Mya and 300 Mya, forests became established as the dominant flora. For most of this period, spore-producing lycopsids, sphenopsids, filicopsids and progymnosperms prevailed. Pteridosperms, fern-like species reproducing by seeds, are known from as early as 370 Mya and are considered to be the ancestors of subsequent spermatophytes, including the angiosperms (Hilton and Bateman 2006). The seed habit permits plants to colonise new habitats and to reproduce without available free water (Bateman and DiMichele 1994), with the consequence that spermatophytes rapidly established themselves in the Permian. By about 260 Mya gymnosperms were on the rise and the giant club-mosses and horsetails in decline. By 200 Mya the conifers had become widely established, including several families with present-day genera such as *Araucaria*, *Pinus* and *Taxus*. My favourite among mid-Permian conifer genera is *Lebowskia*, named for the cult film *The Big Lebowski*. Cindy V. Looy, the taxonomic authority, explains that there is 'a remarkable parallel with the genus here described in that he seems to be a walking anachronism, abiding unnoticed in the "extrabasinal" refugia of modern American society' (Looy 2007). There are, incidentally, further taxonomic connections to the film - in 2006, Agnarsson and Zhang described two new species of African spiders, *Anelosimus biglebowski* and *Anelosimus dude*.

The rise of the seed plants during the Permian, from around 300 Mya, coincided with, and was to a great extent driven by, large tectonic, climatic and environmental shifts. At this time Gondwana and Laurussia had come together to form Pangaea, while globally a period of deep glaciation associated with low atmospheric $CO_2$ concentrations was ending. By the late Permian, $CO_2$ levels were rising and arid environments with fluctuating, generally hot, temperatures were extending (Royer et al. 2004). Under the steamy wetland conditions in which the pre-spermatophyte forests thrived, biodegradation and oxidation processes were suppressed and fossilised remains accumulated in vast quantities as coal and lignite (Taylor et al. 2009). Formation of the Pangaea supercontinent was accompanied by mountain-building and rifting, creating a wide variety of available habitats within which suitably adapted biomes became established (Rees et al. 1999). We see an example of this in modern natural and semi-natural environments where shrubbiness is an increasingly successful trait (Götmark et al. 2016): as described later, new niche opportunities invited exploitation by a range of seed plant forms; in particular, miniaturisation and shorter lifecycles became fitness attributes.

After this journey through the history of the origins and nature of the arborescent life-form, is there an answer to the question: what is a tree? As ever, it depends on who you ask. Lawyers argue about it (for example Goodall 2016). John Evelyn (1664), in one of the first books in English about silviculture, defines a tree thus: 'a lignous woody-plant whose property is for the most part, to grow up and erect itself with a single stem or trunk, of a thick and more compacted substance and bulk, branching forth large and spreading boughs; the whole body and external part, cover'd and invested with a thick rind or *cortex*, more hard and durable than that of other parts; which, with expanding roots, penetrate and fixes them in the earth for stability, (and according to their nature) receive and convey nourishment to the whole'. The *Urban Dictionary* defines 'tree' as 'something that sits in the ground and remains in the same spot for hundreds of years but manages to jump out in front of you on your way home from the pub'. In computer science a tree is a type of hierarchical data structure, and in the case of the Tree of Life Project (Maddison and Schulz 2007), we have an example of matryoshka-like recursion (fittingly for organisms that develop by modular repetition) that places woody plants among the branches of biology's family tree. The present pages have considered trees from the point of view of evolution, architecture, growth and differentiation, and arrived at a kind of archetype, one that serves as the rational counterpart of the Jungian concept of the tree as the psycho-sociological symbol of growth towards self-fulfilment (Jung 1968).

By the Carboniferous, the tree form had arisen independently at least seven times (Boyce et al. 2017), the evolutionary outcome of a transition process that took green plants from the primeval ocean to productive independent life on land. In its underwater environment, a bulky aquatic plant such as kelp is immersed in a medium that supports it physically, nutritionally and physiologically, and protects it from hazards such as harmful solar radiation, atmospheric disturbances and fire. In adapting to the trials and opportunities of dry land, it was more or less inevitable that the first colonisers would evolve into the large, resilient structures we call trees. Size is the essence of a tree. A tree's bulk consists largely of dead cells. Roots and green tissues are constantly renewed. For a tree, elimination of cell contents and of whole cells is a way of life. Trees exemplify the concept of a plant as a population of parts (Harper et al. 1986). This led me to define a tree thus in 1994:

> Plants within a population interact with each other by crowding, altering the light environment, competing for water and nutrients and emitting volatile allelopathic chemicals. The individual leaves on the axis of a single growing plant are also in competition with one another for light, space, materials in the translocation system and hormones and other signalling substances. Because of the fractal, recursive, free-market nature of plant behaviour throughout the hierarchy of biological organisation, it is not at all straightforward to define the demographic unit. Some of the most successful demographic models treat each plant as a population of parts. In this sense, a tree could be considered essentially as a colony of individual autotrophic units living on top of a wooden hill. The units themselves do not need to be individually long-lived for the tree as a whole to achieve great age (Thomas 1994).

The image of a tree as a chimerical association of a lot of dead tissue and a few living, developing cells has echoes of Hagemann's (1999) concept of morphogenesis in the earliest land plants, in which the plant body is divided into a continuously growing 'blastozone' and a steadily decaying 'necrozone'. It's an interesting exercise in extreme reductionism, to condense all the structural and functional factors contributing to organismal unity into the yin and yang of viable cell/dead cell. I've found it helpful in understanding the contrasting life-cycles of trees, perennial herbs and ephemerals (Thomas 2013), which in turn introduces the next episode of our story: the arrival of angiosperms and the rise of the grasses.

**Ginkgo, a living fossil from the age of primeval forest diversification**

*Ginkgo biloba* is the only living representative of the order Ginkgoales, fossilised remains of which date back to the Permian (270 Mya). Ginkgophytes, like other seed plants, probably arose from progymnosperms, but phylogenetic relationships with other early spermatophytes such as cycads, seed ferns and conifers remain unresolved (Gordenko and Broushkin 2015), a situation that may become clearer as a consequence of the recent publication of the draft genome sequence of *G. biloba* (Ruan et al. 2016). *Ginkgo* species expanded rapidly in geographical distribution and diversity from Middle to Late Triassic (240-210 Mya) and were among the most important components of Eurasian Jurassic and Early Cretaceous floras (180-140 Mya). They are believed to have flourished in coastal, swamp or riparian environments that were moist, warm to hot and possibly subject to seasonal drought (Zhou 2009). The period of expansion of ginkgophytes and other woody seed-bearing species extended from the era of primeval pterophytic forests to the dawn of the age of the angiosperms

# PART 2 GRASS

Whispering grass, don't tell the trees
'cause the trees don't need to know

Fred and Doris Fisher *Whispering grass* (1940)

- Mass extinction events, which have restructured the planet's biology at intervals from the Ordovician to the Cretaceous, stimulated plant diversification.
- Angiosperms form a monophyletic group and are estimated to have diverged from ancestor gymnosperms 140 (or perhaps as early as 275) Mya.
- As the angiosperms superseded the gymnosperms to become the dominant flora, there was a transition from woody habit to herbaceous and clonal forms with short lifecycles.
- The trend from trees to herbs is an example of heterochrony and progenesis, involving juvenilisation and precocious reproductive development.
- The evolutionary origins of the flowering plants (Darwin's 'abominable mystery') are complex, with the extinct gymnosperm order Caytoniales currently favoured as the source of the angiosperm lineage.
- The root of the flowering plant phylogenetic tree is divided into the core basal angiosperms, which include the Eudicots and the Monocots, and the so-called ANA sister clade.
- Monocots and dicots diverged around 100 Mya, heading in different developmental directions leading to distinct forms, physiologies and ecologies.
- Monocots and dicots evolved contrasting modes of patterning during embryo, leaf, stem, flower and root development, and of adaptive responses to biotic and abiotic stresses.
- Grasses appeared around 70 Mya and, architecturally and ecologically, established themselves as 'anti-trees', as it were.
- Antagonistic grassland-herbivore co-evolution had a profound influence on grass form, physiology and population structure, and reciprocally on the evolution of grazing animals.
- Ruminants and other ungulates became increasingly abundant and diverse from 50 Mya, when grasslands began to establish as major biomes.
- Grasses adopted silicification and secondary compounds of endophyte origin as defences against herbivory.
- Fire became a critical factor in the balance between grassland and tree-dominated ecosystems.
- Grasses photosynthesising by highly productive $C_4$ pathways became prominent in hot dry regions, succeeding the $C_3$ pathway which is adapted to temperate climates.
- Carbon isotope discrimination is different in $C_3$ and $C_4$ plants, and biomass from the two sources can be identified by the ratio of $^{13}C$ to $^{12}C$.
- Carbon isotope analysis of teeth and collagen allows the diet of fossil herbivores and hominids to be recreated.
- The arrival of the first human ancestors within the last 7 My coincides with the appearance of progenitors of the domesticated grasses.

## Chapter 6 The Boop Principle

The course of evolution has been interrupted - punctuated, to borrow a term from Stephen Jay Gould (Gould and Eldredge 1977) - by explosions in biodiversity and taxonomic extinctions. Based largely on the marine fossil record, five mass extinction events are recognised (Raup 1994): end of the Ordovician (435 Mya), Late Devonian (367 Mya), Permian (245 Mya), Triassic (208 Mya) and Cretaceous (65 Mya). Opinion is growing (Wiens 2016) that we are living through a sixth episode, the Holocene or so-called Anthropocene extinction. Estimates of marine species becoming extinct vary from 76% (Cretaceous and Triassic) to an extraordinary 96% in the Permian event. Floras are particularly resilient and adaptable and were not subject to annihilation on anything like the scale of animal extinctions (Traverse 1988). According to the most recent analyses, plants suffered at most two great extinctions, one spread over 10 My around the Carboniferous-Permian boundary (about 300 Mya), and another in the late Permian (250 Mya), lasting 20 My (Cascales-Miñana and Cleal 2014). Based on abundance of plant fossils, there was an apparent decline of about two-thirds in global genus diversity between 260 and 245 Mya; but this degree of extinction is almost certainly an over-estimate, since changes in geography and climate at this time were unfavourable for fossil preservation (Rees 2002). All in all, it seems that terrestrial vegetation survived episodes of global disruption much better than animals.

Mass extinctions act as a driving force behind evolutionary diversification, restructuring the biosphere on a large scale and increasing the scope for expansion of minor opportunist groups. That an unexploited habitat, however marginal and stressful, represents a niche within which an adaptable organism can make a living is a fundamental rule of evolution and ecology. We left the story of tree evolution with the dominance of the primeval forest, a luxuriant flora comprising hypertrophied vascular plants that had become adapted to the earliest relatively equable swampy environments. The enduring and stable nature of the early forest tree life-form ill equipped it for venturing into uncertain and fluctuating regions of the post-Carboniferous terrestrial environment. Fast-moving opportunist plant types are best fitted to be pathfinders. Accordingly, by 100 Mya, the fossil record shows extensive diversification of form and life cycle.

Rates of molecular evolution and speciation in tall species tend to be lower than in small plants, and extinction rates are negatively correlated with size (Lanfear et al. 2013, Boucher et al. 2017). To keep one step ahead in an unstable world, it generally pays to miniaturise. The rise of angiosperms and herbaceous forms coincided with the increasing variety and insecurity of potential habitats.

Based on (rather ambiguous) fossil evidence, the earliest date for the divergence of the angiosperms is 140 Mya (Wikström et al. 2001, ), though molecular clock estimates imply a much earlier date (Herendeen et al. 2017), perhaps as far back as 275 Mya. The occurrence of both arborescent and non-woody forms in extant lineages, traceable to the earliest fossil angiosperm families, points to the tree life-form as preceding, and the origin of, the herbaceous habit; though the progression from woody to non-woody appears to have been a two-way street (Dulin and Kirchoff 2010, Carlquist 2013). An alternative, though increasingly unlikely, scenario considers the earliest angiosperms to have been rhizomatous aquatics (Soltis et al. 2008).

Except for a few dissenters, most authorities regard the angiosperms to be monophyletic in origin (Soltis et al. 2005). Of the many good reasons for accepting a single common evolutionary source for angiosperms, one is of particular interest because it goes to the heart of what it means to be a plant: the biosynthesis of chlorophyll. Angiosperms need light to make chlorophyll. The light-dependent step in the biosynthetic pathway is catalysed by the enzyme NADP protochlorophyllide oxidoreductase (LPOR). The light required to drive the reaction is captured by a molecular complex of the LPOR protein with its substrates. LPOR is present in all oxygenic chlorophyll-containing organisms. But green algae, pteridophytes and conifers are, almost without exception, also able to make chlorophyll in the absence of light (Armstrong 1998). In these plants the chloroplastic genes *chlL*, *chlN* and *chlB*, structurally unrelated to the nuclear genes for LPOR (Muraki et al. 2010), encode DPOR, a protochlorophyllide reduction pathway that can work in darkness (Bröcker et al. 2010). These genes have been lost from the plastid genomes of all angiosperms, and there is no sign of anything like them in the nuclear genome either. Early in evolution there was clearly a single point at which DPOR vanished without trace from the genome of a common ancestor of all extant angiosperm taxa. The event that eliminated DPOR remains one of the many mysteries of angiosperm origins.

Angiosperms supplanted the gymnosperms beginning in the early Cretaceous, going on to become the dominant flora of modern times. Over this period, several trends favoured the spread of the herbaceous habit. Clonal growth would be an effective strategy for survival in a world recovering from extinction trauma, and the fossil record shows a marked expansion of clonal herbs from the Cretaceous onwards (Tiffney and Niklas 1985). Another adaptive factor would have been the tendency to reach reproductive maturity at a young age (Verdú 2002). The evergreen shrub *Amborella*, the surviving member of a basal clade in the angiosperm lineage, is living evidence that reproductive cycles in early angiosperms were much accelerated compared with those of gymnosperms (Williams 2009). Associated with this ecologically advantageous attribute would be truncation of life-cycles in accordance with the Romano Rule ('live fast, die young'), which enables herbaceous plants to colonise unstable challenging habitats beyond the forest. The developmental relation between tree and herb forms is an example of neoteny (sometimes called heterochrony, paedomorphosis or juvenilisation), a widespread evolutionary principle (Box and Glover 2010). Famously, axolotls are neotenic salamanders. And humans are neotenic apes. The poster child for heterochrony is the cartoon character Betty Boop (Bogin 1999). She is clearly an infant – look at her large head, big eyes, tiny facial features, appealingly cute expression and so forth – but equally clearly she is precociously mature in terms of reproductive characters. Herbs evolved by becoming Betty Boop versions of trees. They retain juvenile characters (for example, trending towards small stature, reduced lignification, ephemeral lifecycle) while attaining early sexual maturity (progenesis). According to this view of plant evolution, broadly speaking, perennial and woody is the foundational plant life-form; herbaceous, ephemeral, annual, biennial and monocarpic phenotypes are reproductively precocious loss-of-function spinoffs (Thomas 2013). Give or take the odd mass extinction, trees and their progenetic, neotenic derivative forms dominated the land flora for tens of millions of years.

## Infantile regression

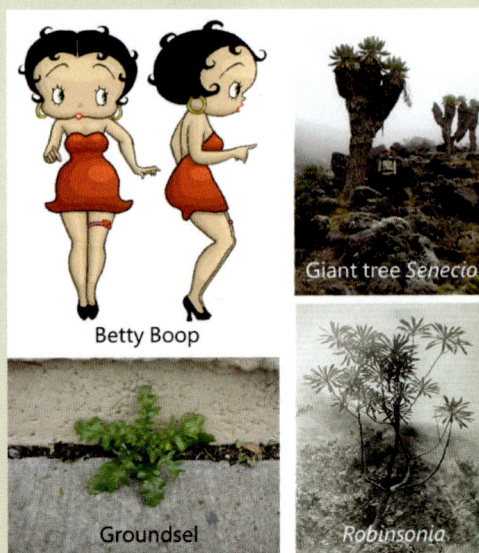

Betty Boop

Giant tree *Senecio*

Groundsel

*Robinsonia*

Betty Boop was a character created in 1930 by cartoonist Max Fleischer, who based her on the popular boop-oop-a-doop Jazz Age flappers Baby Esther Jones and Helen Kane, and triggered much legal disputation for his pains. Betty Boop is the very definition of progenesis - acceleration of sexual maturation relative to the rest of development. So much so that she ran up against the National Legion of Decency and the Production (Hays) Code of 1934 and eventually had to tone down her provocative appearance and behaviour (Fleischer 2005). Within the Asteraceae, the sunflower family, species in the tribe Senecioneae exhibit a wide diversity of morphologies and life-histories, ranging from the small ephemeral herbaceous groundsel to the giant tree types, variously classified in the genus *Senecio* or *Dendrosenecio*. Mabberly (1974) addressed the question as to whether herbaceous habit is derived from the tree form or vice versa. The study preceded the present era of abundant phylogenetic, molecular and fossil evidence, and was primarily aimed at testing EJH Corner's imaginative but now rather discredited Durian theory of tree evolution; but its conclusion is persuasive. In accordance with the Boop Principle, the tree is the more primitive and is the antecedent of the herb form in the groundsels. *Robinsonia*, a genus of tree groundsel sometimes classified with the *Senecio*s, takes its systematic name from Robinson Crusoe. Daniel Defoe based his 1719 novel on the true story of Alexander Selkirk who survived shipwreck and was marooned for four years from 1704 on the Juan Fernández Islands, in the South Pacific Ocean 670 km off the coast of Chile. Forests of *Robinsonia* spp. are endemic to the archipelago and grow nowhere else (Skottsberg 1953). *R. berteroi*, now extinct, is illustrated - the last remaining plant died in May 2004.

## Chapter 7 The abominable mystery

Whenever the evolutionary origins of flowering plants are discussed, it is obligatory to quote from Charles Darwin's letter to Joseph Hooker written on 22 July 1879. The rapid rise and diversification of the angiosperms was, and to a degree remains, a mystery; but it is not entirely clear why Darwin considered the mystery to be 'abominable'. After a scholarly examination of documentation from Darwin, his predecessors, contemporaries and successors, Friedman (2009) concluded that Darwin was troubled by the challenge posed to his gradualist view of evolutionary history. The Darwin-Lyell tradition conceived of evolution as the slow accumulation of small variations and was an explicit rejection of the saltational hypothesis, which allowed for major speciation-scale changes to occur within one or a few generations (Theissen 2009). Perhaps the most dramatic formulation of evolution through the saltation mechanism was proposed by Richard Goldschmidt, who argued that gradual and continuous change is a within-species process, and that for new species to arise there needs to be macromutation and the appearance of rare, successful 'hopeful monsters' (Dietrich 2003). Another view of the gradualness or otherwise of evolution is the notion of punctuated equilibrium (Gould and Eldredge 1977) which, according to Shapiro (2013), is consistent with 'the 21st Century...view [of] the genome as a readwrite...memory system' and is the link between mass extinctions in the fossil record and subsequent bursts of origination.

From an outsider's perspective, it would seem reasonable to accept that different mechanisms, both continuous and discontinuous, will have contributed to evolution, speciation and diversity; but the subject has attracted an astonishing (one might even say abominable) intensity of argument and violent disagreement (see, for example, Sterelny 2007, Pennell et al. 2014, Lieberman and Eldredge 2014). For the purposes of the present book, it isn't necessary to take sides here, and anyway it would be a brave spectator who tried to climb in the ring with such heavyweight combatants as Darwin, Bateson, de Vries, Correns, Goldschmidt, Gould and Dawkins. Suffice it to say that the origins and rapid spread of angiosperms in the Cretaceous remain enigmatic (Hughes 1994).

One of the mysteries surrounding the evolutionary source of the angiosperms concerns the nature of the ancestral seed plant group. Phylogenetic studies of gymnosperms have narrowed down the possibilities to a small number of orders, namely the Glossopteridales, Gnetales, Bennettitales and (currently the most favoured alternative) Caytoniales (Doyle 2012). Of these, only species within the Gnetales have survived extinction, and perhaps the most remarkable of these remarkable survivors is the desert succulent *Welwitschia mirabilis*, which may or may not be a Boop plant (Martens 1977).

In passing from the source gymnosperms to the first angiosperms, we encounter more uncertainty. What groups form the root of the angiosperm phylogenetic tree and how do they relate to their gymnosperm ancestors and to each other? The most basal grade of the angiosperms is the so-called ANA group, named from its three constituent lineages, Amborellales, Nymphaeales, and Austrobaileyales. The Nymphaeales are rhizomatous water-plants and include many species with dramatic flowers. Extant members of the Austrobaileyales are woody plants, the most familiar of which is *Illicium verum*, source of the spice star anise. The order Amborellales has one living member, *Amborella trichopoda*. Fewer than 1% of present-day angiosperm species are descendants of ANA, which is therefore considered to be a sister clade to the main angiosperm lineage (Doyle 2012). The remaining 99% are thought to have radiated from the so-called 'core group' of basal angiosperms (Endress and Doyle 2009), comprising the Chloroanthales, Magnoliids, Ceratophyllales, Eudicots and (the focus of our continuing interest, since it's the grade from which the grasses eventually emerge) Monocots. The phylogenetic relationships of the Chloroanthales, an ancient order which comprises evergreen softwood species that are neither eudicots nor monocots, are uncertain. The Magnoliids is a large group that includes familiar species such as magnolia, avocado and nutmeg. The Ceratophyllales is represented by four extant species of water-plants. The Eudicot clade is the source of all other dicotyledonous species. We shall have a lot more to say about the dicots and monocots in the next chapter.

There has been much speculation about how the basal branches of the angiosperm phylogenetic tree are arranged (Zanis et al. 2002). Founded on anatomical and morphological features, and recent comprehensive analyses of plastid and mitochondrial genome sequences (Drew et al. 2014), a more or less settled picture is emerging in which gymnosperms give rise to two branches. One branch is occupied by *Amborella*, which represents the earliest angiosperm, but which retains the gymnosperm-like character of wood that lacks vessels. The second branch leading from gymnosperms divides in turn into a Nymphaeales branch, comprising aquatic plants lacking a cambium, and a second branch from which are derived the clade represented by *Illicium* and all remaining angiosperms.

Representatives of the basal clades are the centre of attention for enquiries into the origins and diversification of the defining angiosperm structure, the flower (Chanderbali et al. 2016). Flowers are made up of floral structures arranged in concentric whorls, with sepals and petals surrounding the inner, reproductive parts. The models for specification of different floral organs are *Antirrhinum* and *Arabidopsis*, where homeotic mutants that convert one type of floral part into another have been used to dissect the molecular control of flower structure. This established the so-called ABC model and set the conceptual framework for floral morphogenesis in terms of overlapping fields of gene expression (Coen and Myerowitz 1991). For example, class A gene expression is required to make a sepal in the normal position, whorl 1, while A combined with B specifies that a petal will develop in whorl 2. The original three-component scheme has been expanded to include class E genes in the so-called quartet model (Theissen and Meltzer 2007). With one exception, the homeotic A, B, C and E genes encode MADS-box type transcription factors. In the ABCE model, each component gene is expressed in distinct zones of the floral apex as the flower primordium develops, and variants of the model apply to the differentiation of flower parts across all angiosperms.

B and C elements of the quartet model are conserved in specification of gymnosperm reproductive structures. During the evolutionary transition from gymnosperms to basal angiosperms, the quartet function seems to have emerged and diversified through processes of gene duplication, sub- and neo-functionalisation and variations in the stability and degree of overlap of sliding and tight boundaries between the expression domains of floral identity genes (Chanderbali et al. 2016). You can read more about ABCE, as well as floral symmetry, inflorescence development and flower colour in Jones et al. (2013). By the mid-Cretaceous, about 100 Mya, the floral structures of the ANA group, monocots, magnoliids and basal dicots had become established, including the first appearance of attractive tepal-like parts (Sauquet et al. 2017). Thereafter, as the core eudicots diversified and radiated through the late Cretaceous, petal, sepal and nectary structures became included in the floral development program (Friis et al. 2006). Thus the scene was set for grasses to take to the evolutionary stage.

**Flower origins**

Siphonospermum simplex | Grass seed and calyx

Delicate structures like the flowers of early plants don't fossilise well and are rare compared with woody and bony parts, but some beautiful specimens have been recovered. A particularly rich source of lower Cretaceous (127 – 121 Mya; Barrett 2000) flora and fauna is the Yixian Formation of northeast China, which yields assemblages of numerous early seed plants (including species of the basal angiosperm genus *Archaefructus*) as well as many invertebrates, feathered dinosaurs, early birds and placental mammals. *Siphonospermum simplex* from Yixian is a relative of the Gnetales, dating to 125-113 Mya (Rydin and Friis 2010). Gnetophytes are among the gymnosperm progenitors of the basal angiosperms and have flower-like reproductive structures. Fossil records of the floral parts of the earliest grasses are rare. The illustration is of specimens from the Manosque Basin in the south of France, an Oligocene site about 30 million years old (Steur 2016).

## Chapter 8 Anti-trees

Theophrastus (circa 370 BCE) is credited with first recognising differences between monocots and dicots (Isley 2002). Dicotyledones and Monocotyledones were given formal taxonomic standing by John Ray in his *Methodus Plantarum Nova*, published in 1682. Here are the main features by which monocots and dicots are distinguished.

|  | Monocots | Dicots |
|---|---|---|
| **Flower parts** | Multiples of three | Multiples of four or five |
| **Pollen** | Single pore or furrow | Three pores or furrows |
| **Embryo** | Single cotyledon | Two cotyledons |
| **Leaf venation** | Major veins parallel | Major veins reticulated |
| **Stem vasculature** | Vascular bundles scattered | Vascular bundles in a ring |
| **Shoot meristem** | Intercalary | Apical |
| **Roots** | Produced adventitiously | Developed from a radicle |
| **Secondary growth** | Absent | Usually present |

All core dicots (with the exception of a few oddities like the Gunnerales) belong to the Pentapetalae group; that is, their flowers comprise whorls of five structures (Cantino et al. 2007). Monocot flowers, on the other hand, are tri-merous. The distinction between pentamery and trimery appears to be fundamental and exclusive. The many cases of 2- and 4-merous species are considered to be subordinated derivates of the major types (De Craene LR. 2016). The numerical arrangement of floral parts has long been a subject of biomathematical fascination and generally considered to be related to the Fibonacci sequence, which in turn defines the packing of structures at the shoot apex (Endress 1987).

Spores and pollen are essential components of the fossil record (Doyle 2012) and are often present where no other plant structures have been preserved. Their survival over extended periods of extreme geological change is due in large measure to the presence of sporopollenin in their exine layers (shells). Sporopollenin is a remarkable material - to quote the title of a recent review, it is 'the least known yet toughest natural biopolymer' (Mackenzie et al. 2010). Indeed, so chemically refractory is it that it has defied full structural analysis by even the most up to date laboratory procedures. Nevertheless, by characterising mutants in the model species *Arabidopsis* with altered development of anthers and male gametophytes, the biochemical pathways for sporopollenin biosynthesis are beginning to emerge (Kim and Douglas 2013). Pollen assemblages from the early Cretaceous onward have been the centre of attention in attempts to pin down the point of divergence of monocots and dicots, and opinions are divided on what the record is saying (Doyle et al. 2008). There are arguments for 125-100 Mya as the earliest occurrences of pollen fossils with distinctive monocot-type morphologies, and contrary opinions in favour of much later dates - 94-89 Mya have been claimed. For our purposes we can safely settle on somewhere around 100 Mya, give or take a few million years, as the point in which dicots and monocots went their different evolutionary ways. As we have seen in discussing the rise and fall of the Age of the Primeval Forest, 100 Mya is significant as the evolutionary turning point at which plant diversification took off to a degree never seen before or since.

By definition, the feature that distinguishes monocots from dicots must be numbers of cotyledons they possess. But things are not so straightforward. First there's the question as to what, morphologically, developmentally and physiologically, is the definition of a cotyledon. Furthermore, what are the pre-angiosperm evolutionary origins of the cotyledon? And then there is the issue of number - is it really as simple as one for monocots and two for dicots? The lessons of plant evolutionary biology will have prepared us for complications when we address these matters. Traditionally the cotyledon is regarded as a 'seed leaf', that is, the first leaf of the sporophytic phase of the life-cycles of gymnosperms and angiosperms. Another view regards the cotyledon as primarily haustorial (that is, absorbing nutrients from seed reserves to support seedling shoot and root growth) rather than foliar in origin. Sokoloff et al. (2015) have reviewed these types and their evolutionary bases and Burger (1988) has addressed the question as to whether the cotyledons of monocots and dicots are truly homologous structures related to features of the ancestral gymnosperms.

In *Ginkgo* and cycads, cotyledons are fully haustorial. On the other hand most conifers have multiple cotyledons that function as true photosynthetic first seedling leaves. Existing gymnosperm representatives nearest to the basal angiosperms, *Gnetum* and *Welwitschia*, have both haustorium-like structures ('feeders') and cotyledon-like first seedling leaves. Thus the nature and evolutionary relationships of cotyledons remain unclear. There is no significant difference between monocots and dicots up to the globular proembryo stage of early embryogenesis. But soon afterwards, the course of development diverges. The dicot embryo becomes heart-shaped, the precursors of the two cotyledons flanking the central depression that becomes the site of the shoot apex (see Jones et al. 2013). The relationship between cotyledon, shoot axis and apical meristem is more ambiguous in the monocots and probably best understood in the grasses. The haustorial organ of the grass seed (grass fruit, to be more precise) is the scutellum, a lateral appendage attached to the first node of the embryonic shoot. It is a characteristic of monocot embryos that the scutellum appears to be terminal and the shoot apex lateral in the early stages of embryogenesis. The rapidly growing organ appears to be terminal because it displaces the apical meristem sideways. The scutellum may therefore be considered to be the single cotyledon of grasses, albeit one that has no foliar function and is entirely haustorial (Onderdonk and Ketcheson 1972).

Cotyledon morphogenesis is an intrinsic part of the patterning process that establishes the fundamental body plan of the adult plant. By patterning is meant the determination of symmetry, polarity and tissue identity. The globular proembryo in monocots and dicots is radially symmetrical. Thereafter, the monocot embryo becomes bilaterally symmetrical in one plane, whereas the heart-shaped eudicot embryo has two planes of bilateral symmetry. Current models of embryogenesis recognise around 20 genes with defined functions in patterning. Spatial and temporal differences in gene expression will surely have been significant in facilitating the evolutionary divergence of monocots and dicots, but more details of the molecular control of embryogenesis, particularly in monocots, are needed before we know how this might have happened (Zhao et al. 2017).

Like reproductive and embryonic structures, monocot leaf anatomy and morphology are distinctive. The leaves of adult monocots are typically linear and narrow, with vascular tissue arranged in parallel files along the leaf axis. There is no differentiation into blade and petiole; instead the base of the leaf frequently ensheaths the stem. By contrast, the typical dicot leaf has reticulate venation and comprises petiole and lamina (Bell 1991, Scarpella et al. 2004). The contrasting morphologies of monocot and dicot leaves have their origins in patterns of primordium initiation and growth from the shoot apex. The linear, essentially one-dimensional extension of the monocot leaf resembles toothpaste squeezed from a tube, driven by the activity of the intercalary meristem and expansion zone located at the base of the shoot. The dicot leaf generally expands two-dimensionally, growing not only from the base attached to the point of origin in the shoot apical meristem but also from marginal meristems. The source of monocot leaf development and form in angiosperm evolution is uncertain - is linear growth from an intercalary meristem ancestral or is it derived from a more ancient eudicot-type? On balance, the former alternative seems to be favoured (Rudall and Buzgo 2004). The leaves of angiosperms are better fitted than those of extant and fossil gymnosperms and ferns for the primary functions of supplying water across the photosynthetic surface and reconciling the need to manage evaporative demand with keeping stomata open to allow access to atmospheric $CO_2$. The evolution of optimised foliar structure and function in angiosperms contributed to a step-change in productivity and diversification of ecological range in early dicots and monocots (Zwieniecki and Boyce 2014, Boyce et al. 2017).

In many monocots internodes are compressed and the sheaths of successive leaves form a tight concentric pseudostem. Leaf bases are often swollen and function as perennating or storage organs (see Bell 1991, Jones et al. 2013). Vascular bundles in cross-sections of monocot shoots are generally scattered, unlike those of dicots, in which the cylindrical arrangement of the vasculature divides the stem into cortex and stele (Esau 1960). Radial growth of tree trunks and other dicot stems is supported by the activity of the lateral vascular cambium. Most monocots are herbaceous and lack such a cambium, but arborescent monocots such as palms and bamboos produce woody stems by anomalous secondary growth (Tomlinson and Zimmermann 1969). It is generally considered that possession of vascular cambium and the capacity for secondary growth is the primitive condition in basal angiosperms and that competence for secondary growth was lost on divergence of the monocot lineage. Anomalous secondary growth and cambium in monocot stems are not homologues of dicot functions and structures (Spicer and Groover 2010).

Grasses emerged rather late in monocot evolution (Willis and McElwain 2014). The anatomical, morphological and physiological features of the monocot body-plan were seemingly well established around 100 Mya. And yet the evidence points to no earlier than about 70 Mya for the appearance of the grasses (Poaceae). Grasses accumulate large quantities of silica, much of which is deposited in the outer walls of leaf epidermal cells (Kaufman et al. 1985). Phytoliths are the fossil residues of silica cells and their species-specific morphologies make them diagnostic for palaeological analysis. The oldest confirmed fossils of Poaceae are phytoliths of *Changii indicum*, a progenitor of the rice tribe, recovered from dinosaur coprolites dated to 66 Mya (Prasad et al. 2011). A number of factors contributed to the delayed establishment and subsequent rapid expansion of grassland ecosystems (Strömberg 2011). Although grasses were ancestrally adapted to warm, relatively mesic, closed environments, their morphological and physiological flexibility equipped them rapidly to exploit new, often drier open habitats as climate and geography changed during the Cenozoic era (from about 65 Mya). The development of high-productivity, water-efficient $C_4$ photosynthesis around the beginning of the Oligocene was decisive for expansion into tropical-subtropical lowlands and regions with warm-season precipitation (Edwards et al. 2010, Edwards and Smith 2010). Another stimulus was the coevolutionary relationship between grasses and grazing animals; and a major force shaping the nature and extent of grassland biomes was, and is, fire (Bond and Keeley 2005, Pennington and Hughes).

The first campaigns in the War of this book's title saw the opening-up of vegetation and expansion of $C_3$ and later $C_4$ grasses occurring at the same time that tree-dominated biomes were in retreat. Climate, fire and herbivory all played a part in altering the balance between woodland and grassland (Bond 2008). Grasses became successful precisely because they evolved to be intrinsically non-treelike: anti-trees, one might say. Contrasting life-forms are emphasised in the Raunkiær (1937) system. In this scheme, woody perennials (trees and shrubs) are termed *phanerophytes* and their characteristic feature is the position of resting buds, more than 0.5 m above ground. Shrubs and other small woody perennials with resting buds at a height of less than 0.5 m are *chamaephytes*. Grasses are classified as *hemicryptophytes*, herbaceous perennial plants with the resting bud at or near the soil line. Perennial herbaceous species where the resting bud is below ground are *cryptophytes*; if their shoots or roots are modified to be storage organs, such as bulbs, corms, rhizomes and tubers, they are *geophytes*. The position of resting buds relative to the surface of the ground is significant for the resilience of an individual or community in the face of environmental challenge. Phanerophytes and chamaephytes, which dominate woodland biomes, are vulnerable to the depredations of temperature and water stress, fire, grazing and browsing. By contrast the (hemi)cryptophytes of grasslands and savannahs are physically configured to avoid, and recover from, such environmental challenges. To be an anti-tree is to have a critical advantage in the evolutionary arms-race during the Cenozoic era.

45

## Tree and anti-tree

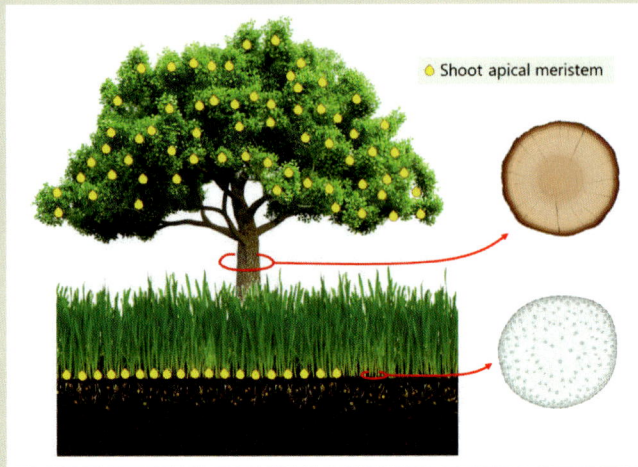

Comparing the tree form with that of grasses reveals several fundamental differences with profound implications for adaptation and survival. Grasses keep their growing points close to the ground, whereas those of trees are exposed at height. Grasses have abandoned cambium and secondary xylem whereas the essence of a tree is its woody frame. The two life-forms have adopted contrasting strategies for storing reserves in vegetative tissues. Evergreens tend to have low rates of internal nutrient cycling, and deciduous trees withdraw nutrients to the bark before shedding foliage. Grasses, on the other hand, accumulate reserves in leaf bases or underground organs such as stolons, rhizomes and tubers (see Jones et al. 2013). Trees and grasses have very different root systems, and rhizosphere interactions are at least as important for tree-grass competition as herbivory, shading and other stresses in the aerial environment (Dohn et al. 2013, Van Noordwijk et al 2015). It's increasingly recognised that below-ground adaptations and processes that are hidden from sight are of global significance for plant evolution and biosphere function (Bardgett et al. 2014). Differentiation into root and shoot was one of the earliest capacities developed by the first land plants, and had profound geochemical consequences (Raven and Edwards 2001). There are indications of fungal (paramycorrhizal) associations in Rhynie Chert fossils from the lower Devonian, and for arbuscular-type eumycorrhiza in the earliest forest ecosystems (Kenrick and Strullu-Derrien 2014). Plant-mycorrhiza symbioses are almost universal and are not only essential for host fitness through the integrative function of hyphal networks but also critical for cycling of carbon, nitrogen and phosphorus in ecosystems (Heijden et al. 2015).

## Chapter 9 Forage and fire

Comparing the tree form with that of grasses has revealed several fundamental differences with profound implications for adaptation and survival. Now we look in more detail at the environmental context for the long-range competitive interaction between forest and grassland ecosystems (Wilson 1998, Baudena et al. 2015), beginning with biotic interactions. Grasses co-evolved with grazing and browsing animals, such that there was reciprocal influence on the development of structural and behavioural traits (Stebbins 1981). Interaction with grasses drove the evolution of herbivore body size, herd behaviour, specialised teeth and the development of the rumen and hindgut to ferment fibrous biomass. For their part, grasses responded to animals by adopting intercalary (that is, ground-level) growth, distinctive foliar architecture and tolerance of defoliation and trampling. Grazing and browsing favour grasses by limiting tree establishment and opening up closed forest systems. It's important to note, however, that the term 'grassland' denotes what Beerling (2007) calls a 'cosmopolitan biome' - an ecosystem that includes open mixed communities such as savannah in which grasses are not the sole, or even majority, plant type.

At the start of the Cenozoic era (65 Mya) the supercontinent Pangaea had broken up into the continental landmasses with more or less the shapes and locations we know today. The major biomes – tropical, subtropical and temperate – were still dominated by woody plants, but angiosperms were increasingly abundant (Willis and McElwain 2014). The earliest known biomes dominated by open-habitat $C_3$ grasses with associated fauna became established in South America between 40 and 35 Mya. $C_3$-dominated grassland systems appeared elsewhere later, in the case of South-East Asia and Australia as recently as 10 to 5 Mya (Strömberg 2011). The rise of the $C_4$ grasses took place between 18 and 12 Mya (Edwards et al. 2010). What animals were around to exploit the productivity of these ecosystems at this time? One discovery that made the headlines, as often happens where these creatures are concerned, is that dinosaurs might have overlapped with, and dined on, the earliest grasses (Piperno and Sues 2005); but they couldn't have done so for long - the catastrophic event 66 Mya (probably a meteor impact as proposed by Luis Alvarez and colleagues; Wohl 2007) that wiped out the dinosaurs, and much life on earth, predated and indirectly stimulated expansion of grasslands in the Cenozoic era. Nevertheless, the diet of herbivorous dinosaurs raises interesting issues concerning adaptations of the eating and digestive systems, and the origins of gigantism, in the early fauna of grasslands (Clauss et al. 2013, Strömberg et al. 2016).

Following the demise of herbivorous dinosaurs, we enter the age of grazing and browsing mammals. The hoofed mammals are broadly divided into the odd-toed (Perissodactyla) and even-toed (Artiodactyla) ungulates. The odd-toes include the equids, tapiroids and rhinoceroids. All are hind-gut rather than rumen fermenters. The even-toes include the ruminants (cattle, sheep, goats, giraffes and deer) as well as the non-ruminant suiformes (pigs) and camelids (Janis 2007). The rise of the ungulates, which took off around 50 Mya, tracked the expansion of grasslands. The herbage which supported the expansion of grassland fauna is generally considered to be a foodstuff of inferior quality, because of its high content of slowly digestible or indigestible fibre, low levels of protein and the presence of antinutritional secondary plant metabolites such as tannins. The digestive system, and associated morphological, physiological and behavioural adaptations, takes a diversity of forms across the range of ruminants (Hofmann 1989). Gigantism, a feature of non-ruminant fossil herbivores – dinosaurs, *Paraceratherium* (the giant rhinoceros of the late Oligocene – Prothero 2013) and mammoths, for example - is thought to be the outcome of complex ecological and allometric relationships between forage quality, biomass availability and the high passage time for digesta made possible by hindgut, as opposed to rumen, fermentation (Clauss et al. 2013).

Teeth and phytoliths survive fossilisation in a state that allows reconstruction of the adaptive measures and counter-measures adopted by grasses and the grazers that prey on them. Ungulate dentition is generally hypsodont, that is, it comprises ever-growing high-crowned teeth with enamel which extends past the gum line. The muzzle in grazers tends to be broad, whereas that of browsers is narrow (Janis 2007). The facile interpretation of the evolutionary origin of foliar silicification is that it is a herbivore deterrent. According to this view, the highly hypsodont cheek teeth of grazers is an antagonistic coadaptation to an abrasive diet. Critical examination of the evidence reveals a more complex picture (Strömberg et al. 2016). Silica accumulation, which appeared independently numerous times during plant evolution, may, for example, have originated for structural support as a low-cost alternative to lignin, or to limit evapotranspiration. Moreover, vertebrates were far from being the first or only herbivores that would have been deterred by silicification (Hartley and DeGabriel 2016). Nevertheless, it is reasonable to consider hypsodonty in ungulates to be an adaptive response to the problem of constant dental wear that comes with almost continuous intake and mastication of gritty, fibrous plant material with low nutrient content. The relationship between grazing and the accumulation of silica in the grassland diet is of global significance. The expansion of grasses greatly increased the input of silicates into the biosphere and, via the grazers and on through the food-chain, delivered this formerly scarce mineral to the oceans. There it stimulated the evolution of diatomaceous phytoplankton and productivity of marine ecosystems (Falkowski et al. 2004).

Despite the beliefs of some devotees of 'natural food', no plant *wants* to be eaten, though it might strike a bargain with a predator on the basis of cost-benefit balance. In this context, a potent survival weapon is a chemical one: plants are prodigious fabricators and accumulators of bioactive secondary compounds (Harbourne 2014). Grasses are no exception, having the capacity to make hydroxamic acids, condensed tannins, cyanogenic glycosides and other chemical deterrents. But they do not match the dicots in the quantity and variety of toxins and psychoactives they accrue (Vicari and Bazely 1993, Jones et al. 2013), relying as much on tolerance of physical damage by herbivores and pathogens as on biochemical retaliation. The relatively benign chemical composition of grasses is one reason why they feed the world. Nevertheless, from the earliest times in evolution, plants in general, and grasses in particular, have enhanced their arsenals of secondary compounds by forming mutualistic associations with fungal and other sources of bioactives (Saikkonen et al. 2013). A remarkable recent fossil discovery is of an ergot-infected grass inflorescence embedded in amber that dates to 100 Mya, at the dawn of monocot evolution (Poinar et al. 2015). Ergot, the external fruiting body of species in the fungal family Clavicipitaceae, is notorious for its extreme psychotoxicity when it gets into the food chain. And thereby hangs a tale, which will be briefly sketched (because it's a particular interest of mine).

Under some circumstances, present-day temperate pasture grasses of the genera *Lolium*, *Festuca* and *Schedonorus* elicit toxic responses in livestock (Siegel et al.

1987). The cause is a brew of bioactive compounds produced by endophytes. An endophyte is a bacterium, actinomycete, mycoplasma or, as in the case of grasses, a fungus, that lives within a plant for at least part of its lifecycle. The relationship of host to fungal endophyte extends beyond the mutualistic provision of chemical defences in exchange for a home. The *Lolium* endophyte has been shown to exert an appreciable influence over the host's development and stress tolerance through genetic reprogramming (Dupont et al. 2015). The major endophytes of *Lolium* are classified in the genus *Epichloë* (Leuchtmann et al. 2014). Association with endophyte is systemic and the asexual epichloae are passed from host generation to generation almost exclusively by vertical transmission. At least four types of bioactive compound are produced by the endophytes of *Lolium*: lolines, indole diterpenes, ergot alkaloids and peramine (Schardl et al. 2012). Inclusion of ergot bioactives in this list is significant: the genus *Epichloë* is within the ergot family Clavicipitaceae (Kuldau et al. 1997) and the record of toxicity attributed to endophyte has become well and truly scrambled with ergotism (Thomas et al. 2011). Not only does it have implications for the co-evolution of grasses and grazers, endophyte plays a leading part in a central theme of human cultural history, to which we will return in Chapter 13.

As well as herbivores and pathogens, abiotic stresses - short term, seasonal and climatic – invoke acclimatory and adaptive change in vegetation systems. In this respect, fire has been of particular significance since the earliest period of terrestrial life (Beerling 2007, Looy 2013). The plants of inflammable ecosystems have a range of adaptations that allow them to tolerate or avoid regular fires, and in many cases periodic burning is a life cycle requirement. For example, the seeds of many species from fire-prone regions are dormant until they are exposed to karrikins, a class of growth promoting hormone present in smoke from burning vegetation (Flematti et al. 2004). Some trees are adapted to, and even benefit from, periodic burning (Rowe 1983); but generally fire is a necessary and recurrent factor in many natural environments, and serves to favour grasses by limiting the extent of forests. Fire is lethal for most chamaephytic and phanerophytic woody plants, but because of the way they are made, grasses and other (hemi)cryptophytes and geophytes can survive burning to regrow. Fire became a decisive factor in shaping the balance between grassland and forest across much of the Earth (Scheiter et al. 2012). Regular burning is a natural feature of the world's grasslands and savannahs to this day. Vegetation models predict that, without fire, vast areas of African and South American grasslands and savannahs would potentially form forests under current climate conditions, and worldwide there would be a doubling of forested areas (Bond et al. 2005). The severe selective pressure exerted by fire might explain the evolution of the extraordinary savannah and cerrado growth forms known as *geoxyles* - underground trees (Maurin et al. 2014). The geoxyle habit, which appeared independently several times during the last 2 My, consists of a subterranean system of woody roots and shoots (xylopodia) with a high capacity for resprouting and producing new shoot buds. We might regard geoxyles as trees that have abandoned the chamaephytic/phanerophytic lifestyle with its vulnerability to burning and adopted the safer habit of cryptophytes and geophytes. A case of if you can't beat them, join them.

**Grazing and burning**

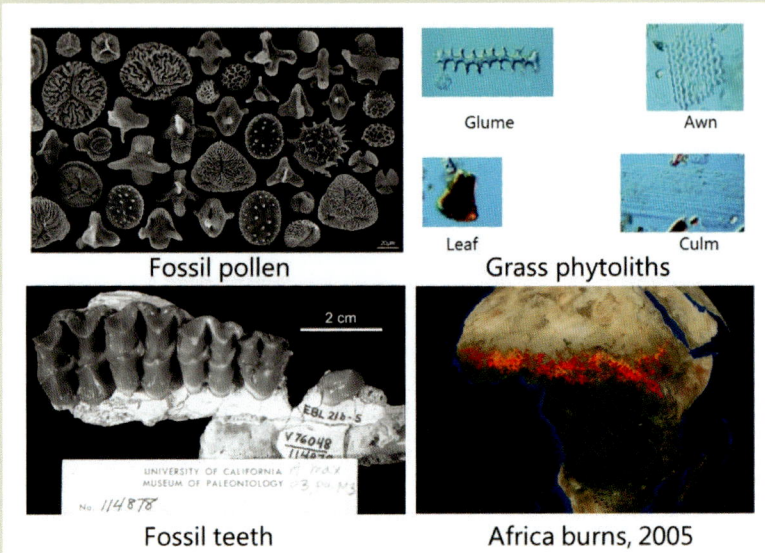

Fossil pollen

Grass phytoliths

Glume

Awn

Leaf

Culm

2 cm

Fossil teeth

Africa burns, 2005

Pollen, phytoliths and teeth are the most persistent and informative fossil indicators of the evolutionary history of grassland. The illustration shows the remarkable state of preservation of a pollen assemblage from 66 Mya, found in bedrock in south western North Dakota, USA. Fossil phytoliths take many distinctive forms, which allow them to be assigned to organs, orders and, in some cases, families of grasses: see Ghosh et al. (2011). The diet of ungulate grazers of the first grasslands is reflected in their characteristic dentition. Oreodonts are among the most commonly represented hoofed mammals in the North American fossil record between 47 and 7 Mya. The eminent palaeontologist Joseph Leidy (1823-1891) referred to oreodonts as 'ruminating hogs'. The illustration is a specimen of oreodont cheek teeth, which are similar to those of ruminant artiodactyls, although their front teeth are more like those of pigs and peccaries. Several lineages of oreodonts are thought to have developed hypsodont teeth independently under selective pressure to adapt to a grazing diet (Mihlbachler and Solounias 2006). Fire behaves, in the memorable phrase of Bond and Keeley (2005), like a 'global herbivore'. Satellite imaging by the Moderate Resolution Imaging Spectroradiometer (MODIS) follows the burning season in Africa, which begins early in the year with widespread fires across the Sahel and savannas and subsequently moves south until Southern Africa is ablaze by early July. Fire is a normal part of the annual cycle in many landscapes worldwide, but is also used by people to create and maintain agricultural areas, often with harmful consequences for the natural and inhabited environment (McWethy et al. 2013).

## Chapter 10 First contact

In *Leaves of Grass* (1892), Walt Whitman wrote 'I guess the grass is itself a child, the produced babe of the vegetation'; in his entertaining book on the nature of crops, my friend John Warren states that 'many people regard grasses as being rather dull' (Warren 2015). But whether dull or babe, grasses are essential for life on Earth, with an influence extending from the heart of human civilisation to the depths of the ocean. From their first, almost discreet, entrance onto the world stage around 70–100 Mya, the Poaceae have proliferated to become leading players in the biosphere today, comprising 650-785 genera and around 10,000 species, making them the fifth-largest plant family. Their success reflects their uniquely wide adaptability under conditions of warm or cool seasons. Grasses can be annual or perennial (but there are no biennials) and range in size from minute Boop species such as *Brachypodium* (a favoured model grass for molecular genetics research – Catalan et al. 2014) to large tree bamboos (Farrelly 1984). Grasslands in the broad sense account for 40% of terrestrial surface vegetation outside polar regions, and as much as 60% of net primary productivity. On the whole, diversity of monocots as producers and vertebrates as consumers are correlated, but the extent to which this is a direct relationship or a consequence of collinear responses to environmental gradients is difficult to determine (McInnes et al. 2013).

Most grass species, including those of major significance for the subject of this book, belong to one of two clades of the family Poaceae. The BEP (sometimes called BOP) grouping is named for the subfamilies it comprises: Bambusoideae (bamboos), Ehrhartoideae (or Oryzoideae, includes rice) and Pooideae (temperate grasses). The PACMAD clade includes the panicoid grasses and groups with C$_4$ photosynthesis. Maize, sorghum and sugarcane are in this complex, which takes its name from the constituent subfamilies Panicoideae, Arundinoideae, Chloridoideae, Micrairoideae, Aristidoideae, and Danthonioideae. It is proving difficult to calibrate the timeline for the appearance of the BEP-PACMAD crown Poaceae clades and their and subsequent split (Christin et al. 2014), but these divergences seem to have happened around 60 Mya give or take a few million years (Jones et al. 2014). This uncertainty is frustrating because it makes it difficult to identify which ecological, climatic and geochemical factors drove dispersal of lineages and the appearance of C4 photosynthesis. Nevertheless, it is pretty safe to picture the world of 10 Mya in which C4 grassland species begin their expansion (Strömberg 2011) while tree and shrub-dominated biomes cover much of the land outside polar regions, steppe, open savannah and semi-desert (Willis and McElwain 2014). But things are about to get interesting: hominids enter the scene.

With the exception of rice (*Oryza*), ancestors of which may date back to 25 Mya or even tens of My earlier (Prasad et al. 2011), the progenitors of the major present-day cereals and pasture grasses appeared around the start of the Pliocene epoch. They include the genera *Triticum* (ancestor of wheat), *Hordeum* (barley), *Avena* (oat), *Secale* (rye) and *Lolium* (ryegrass), all members of the Hordeae and the Aveneae/Poeae tribe complex within the Poideae (Hochbach et al. 2015). Molecular phylogenies constructed for the C4 panicoid grasses are consistent with diversification of the antecedents of maize (*Zea*), *Sorghum* and sugarcane (*Saccharum*) at around the same period (Figueira et al. 2008, Bouchenak-Khelladi et al. 2010, Jones et al. 2014). On a similar timeline, the first hominins began to appear. The human ancestral lineage became distinct from the New World Monkeys 57.5 Mya ago, the Old World Monkeys 31 Mya ago, the gorilla 8.1 Mya ago and the chimpanzee 4.5 Mya ago (Takahata and Satta 1997). *Sahelanthropus*, a fossil from Chad, dated to 7 Mya, may be the earliest specimen of a hominid (Brunet et al. 2002). *Australopithecus* (including the famous 'Lucy') appeared around 3 Mya (Antón et al. 2014). To be clear when discussing human evolution, it's worth mentioning the distinction between the terms 'hominin' and 'hominid'. Hominid refers to the group consisting of all modern and extinct great apes and their ancestors, and includes modern humans, chimpanzees, gorillas and orang-utans. The hominins comprise modern, extinct and immediate ancestral humans, including members of the genera *Homo* and *Australopithecus*.

It is estimated that the earliest use of tools did not occur before 2.6 Mya. For all hominins subsequent to about 3.5 Mya isotopic studies identify a diverse diet incorporating a broad range of C3 and C4 plants (Cerling et al. 2013, Sponheimer et al. 2013). Humans and grasses had finally met, and their fates have been intertwined ever since.

## Carbon isotope analysis: you are what you eat

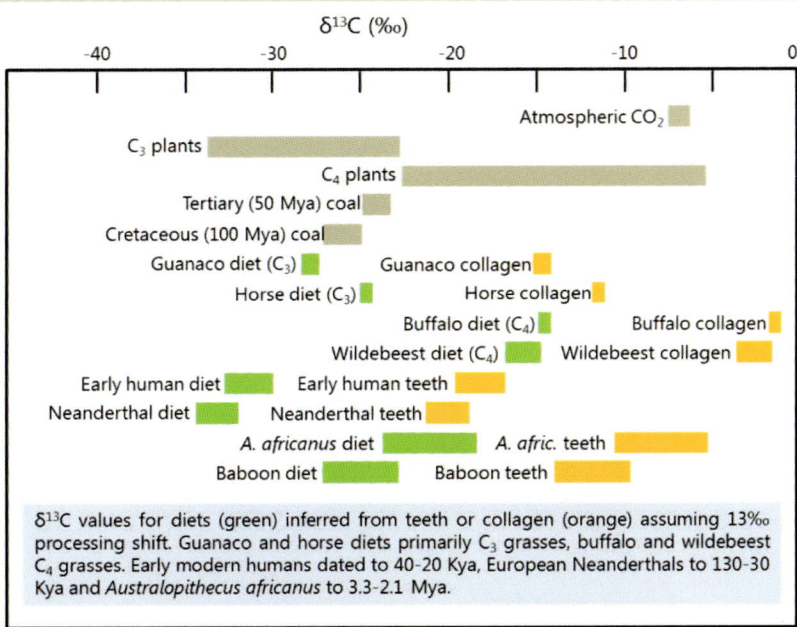

$\delta^{13}C$ (‰)

-40    -30    -20    -10    0

Atmospheric $CO_2$

$C_3$ plants

$C_4$ plants

Tertiary (50 Mya) coal

Cretaceous (100 Mya) coal

Guanaco diet ($C_3$)    Guanaco collagen

Horse diet ($C_3$)    Horse collagen

Buffalo diet ($C_4$)    Buffalo collagen

Wildebeest diet ($C_4$)    Wildebeest collagen

Early human diet    Early human teeth

Neanderthal diet    Neanderthal teeth

A. africanus diet    A. afric. teeth

Baboon diet    Baboon teeth

$\delta^{13}C$ values for diets (green) inferred from teeth or collagen (orange) assuming 13‰ processing shift. Guanaco and horse diets primarily $C_3$ grasses, buffalo and wildebeest $C_4$ grasses. Early modern humans dated to 40-20 Kya, European Neanderthals to 130-30 Kya and *Australopithecus africanus* to 3.3-2.1 Mya.

The core light-driven and $CO_2$-fixing processes of photosynthesis – the $C_3$ pathway - are almost as ancient as life on Earth, but land plants have developed enhancements that allow them to function in stressful (particularly water-limited and hot) environments (see Jones et al. 2013). The $C_4$ variant evolved several times independently in grasses and many other plant groups of tropical and subtropical provenance. Carbon fixed by the $C_3$ and $C_4$ routes makes its way into the plant and on through the food chain. The carbon of atmospheric $CO_2$ exists as two stable isotopic forms, $^{12}C$ and $^{13}C$, in the approximate ratio 98.8:1.1. It's a property of the active sites of the primary carbon-fixing enzymes of $C_3$ and $C_4$ plants (rubisco and PEP carboxylase) that they handle $^{12}CO_2$ and $^{13}CO_2$ differentially, resulting in a distinctive $^{12}C$:$^{13}C$ signature in respective biomass and food chain compositions (Farquhar et al. 1989). Isotopic ratio is generally expressed as $\delta^{13}C$ (‰). Organic material derived from $C_4$ photosynthesis is relatively enriched in $^{13}C$. The $\delta^{13}C$ value of $C_3$ vegetation is in the range -23 to -33‰. Fractionation between consumption by a herbivore and incorporation into tooth enamel or collagen increases the $\delta^{13}C$ by about 12 to 14‰ (Lee-Thorp et al. 1989, Cerling and Harris 1999). The teeth or collagen of animals that eat $C_3$ vegetation have $\delta^{13}C$ values between about –10 and –16‰ whereas consumers of $C_4$ tropical grasses have $\delta^{13}C$ values between 2 and –2‰ (Cerling and Harris 1999).

## PART 3 HUMAN

"Give me the plains and the lakes!" he thought. "There you can see what you are likely to meet. Now if this were a grove of little birches, it would be well enough, for then the ground would be almost bare; but how people can like these wild, pathless forests is incomprehensible to me. If I owned this land I would chop down every tree."

Selma Lagerlöf *The Wonderful Adventures of Nils* (1907, translated by Velma Swanson Howard)

- Observations on tool use by the primate relatives and ancestors of the genus *Homo* are informative about the origins of human-plant relationships.
- From 2 Mya, plant USOs (underground starch-storing organs) were accessed and made palatable through the application of digging tools and cookery.
- The high-glycaemic USO-based diet profoundly influenced hominin physiology (including brain size), behaviour, culture and expanding geographical range.
- The geological period over which humans co-evolved with the plants that meet their nutritional and energy needs is often called the Anthropocene.
- At the dawn of agriculture around 10 Kya, the change from USOs to cereal seeds as sources of starch resulted in a population explosion that has continued.
- Across the world, civilisations emerged, based on plant and animal domestication, and initiated an ongoing era of deforestation.
- Hunting and domesticating grazing animals made the large source of dietary energy in cellulose accessible, further powering social advance.
- For ancient cultures, forests were savage, forbidden places with special significance in myth, ritual, religion and the law.
- The Fall of Man was often represented as estrangement from a perilous forest paradise and condemnation to a life of toil in wheatfield and pasture.
- Weeds exploited the spread of domesticated grasses and other field crops from centres of agricultural civilisation.
- Invasion of arable crops by interloper weeds from pastures introduces a new layer of complexity into the evolutionary tension within and between grasses and trees.
- Fields and forests are the settings for Biblical and Shakespearian accounts of hubris and madness.
- Forest clearance occurred not just to make way for agriculture but also as acts of war, desecration or religious observance.
- Cultivation and exploitation of some tree species has been practiced for millennia, and trees became assimilated into Arcadian landscapes in the post-Christian era.
- Romanticism was associated with trees as symbols of political conservatism and an introspective reverence for forests and wild places, a frame of mind that persists in the present-day New Nature Writing vogue.
- Nevertheless, human dependence on grasses for food, feed and recreation remains as strong as ever and, in *The War Between Trees and Grasses*, there is no doubt whose side *Homo sapiens* is on.

## Chapter 11 Meet the ancestors

In previous chapters I've looked in some detail at the evolutionary origins of trees and grasses. Now humans come into the picture, but for good reasons I do not intend to pay the same degree of attention to *Homo sapiens* and its precursors. First, I'm a plant scientist and have enough biological understanding to give a defensible account of the progression from emergence onto land through the rise of the primeval forest to the appearance of angiosperms and eventually the grasses. That's not to say that colleagues much more expert than I might take issue with my view, but at least disputation would be on more or less equal terms. The issue of human origins, however, is a different kettle of fish (to use a quite inappropriate expression). It's one thing to talk broadly about whole phyla, such as gymnosperms or monocots, and another to narrow the focus to a single species. Knowledge of human origins is in a state of constant reconsideration as new fossil remains are discovered and new techniques are introduced (such as analysis of ancient DNA). Self-interest as a species (perhaps narcissism is a better term) intensifies the scrutiny to which each new revision of human evolution is subject, leading sometimes to vehement disagreement, with which I have no wish to engage. I keep in mind the experiences of the late Elaine Morgan, a distinguished writer and playwright (and Welsh, what's more), who promoted the idea, first proposed by the marine biologist Alister Hardy, of the 'Aquatic Ape' origin of *Homo sapiens*. Her arguments encountered much criticism from the mainstream (for example Langdon 1997, Kuliukas 2014) and even hostility bordering on trolling, at least in part because she was a non-scientist with a feminist socio-political agenda. It seems that resistance to the theory is beginning to soften somewhat, but there's a lesson here that I take, about steering clear of trouble you don't need.

Nevertheless, to understand those aspects of human origins that speak of the formative relationships with trees and grasses, it is instructive, and perhaps not too controversial, to look at the behaviour of our primate relatives. In this connection, the use of tools is of particular significance. Evidence of tool-making by hominins in the period 2.6 to 1.7 Mya (the so-called Oldowan industry) is widespread across archaeological sites in Africa, Europe, the Middle East and South Asia (Semaw 2000). Even older records (Lomekwian tools) from a site in

Kenya have been dated to 3.3 Mya (Harmand et al. 2015). The use and fabrication of enabling implements is widespread and diverse in modern primates (and other animals too). Among the Old World apes and monkeys, more than 80 distinct uses of tools in the broad sense have been observed. Comparison of Oldowan technology with the tool-using capabilities of modern chimpanzees led Wynn et al. (2011) to conclude that 'in its general features Oldowan culture was ape, not human'. It is reasonable, therefore, to infer the uses of tools by early humans from the activities of extant primates.

To accomplish a variety of tasks, chimpanzees and other Old World apes and monkeys employ a range of materials, including stone and wooden hammers, anvils and cleavers, probes and spears and digging sticks (Wynn et al. 2011, McGrew 2013). The most frequent use of tools by chimpanzees living in arid environments similar to those reconstructed for early hominins is for feeding (McGrew 1992, Hernandez-Aguilar et al. 2007). Chimpanzee behaviour and diet are considered to provide insights into adaptations of early hominins (Hernandez-Aguilar et al. 2007, Hernandez-Aguilar 2009). A significant feature of the diverse diet of chimpanzees is the inclusion of out-of-sight foodstuffs, obtained through what Parker and Gibson (1977) term 'extractive foraging'. By employing digging tools to access such sources of food, wild chimpanzees and capuchins are displaying degrees of intelligence, technological innovation and physical dexterity that approach those of humans (Laden and Wrangham 2005).

Chief among the objects of this digging activity are the underground storage organs of plants, generally abbreviated as USOs (Wynn et al. 2011). USOs, which include roots, bulbs, tubers and rhizomes, are regarded as nutritional sources that had a critical role in the adaptive development of hominin physiology, behaviour and culture (Hardy et al. 2015); and of course they are characteristic of the savannah and other grassland environments within which early human evolution took place. Chimpanzees are often observed manually extracting USOs without the requirement for a tool where the ground allows, but Wynn et al. (2011) argue that in general they are animals that are dependent on the use of tools to function adaptively in their environment. The USO zone in grasslands can be a challenge for even the toughest implements. The story is told of the pioneering homesteaders in the prairies of North America whose iron ploughs were broken by the soils and mats of native grasses subjected to age-long compaction by herds of grazing bison, a frustration that led directly to the adoption of John Deere's 'Sodbuster' steel plough in the 1840s (Friedman 2015). Thus we see, and not for the last time, recapitulation of an encounter between hominid and grass directly driving innovation in technology and behaviour.

What was the attraction of USOs for hominins? The answer, in a word, was starch. Starch constitutes up to 80% of the dry weight of USOs. And what's so special about starch? For one thing, it's a highly compact source of stored bioenergy. Mole for mole, organic carbon in the form of starch is more than 5 times more concentrated than it is in a sucrose solution isosmotic with seawater (Raven 2005). The capacity to lay down starch in perennating and resting structures is one of the distinctive traits evolved by herbaceous species. It enables fluctuations in the supply of, and demand for, resources necessary for plant growth to be evened out. Raw materials accumulated during periods of abundant carbon and nutrient assimilation represent an investment against times when the environment becomes unfavourable. Cells packed with starch and other osmotically inactive storage polymers, which to a significant degree also displace intracellular water, are intrinsically resistant to dehydrating stresses. Mass mobilisation of reserves powers rapid growth and development, giving the organs and individuals of the next generation of plants a competitive edge (Raven 2005, Jones et al. 2013). Hardy et al. (2015) point out that USOs can be dried to increase durability and portability, or left undisturbed in the ground, where they remain stable and can be harvested as needed over a period of months. The distinctive features of starchy USOs had profound implications for the direction of human evolution.

A starch-rich diet is thought to have permitted Australopithecines to expand into new habitats and to have led to a reduction in the length of the digestive tract. Moreover, consumption of USOs may have influenced the structure of the teeth and jaws of early hominins (Laden and Wrangham 2005). It's conceivable that foraging for USOs in liminal shallow water habitats was a factor in the evolution of bipedal gait (Wrangham 2009), a hypothesis that might effect a partial reconciliation between the savannah and aquatic/waterside ape narratives of human origins. USO consumption in early *Homo* social groupings has been associated with 'grandmothering', the undertaking of foraging and sharing tasks by post-menopausal females, thereby releasing younger females for more frequent childbearing and rearing (O'Connell et al. 1999). USOs are classified as 'fallback foods', that is, lower quality foods whose exploitation to minimise wasting and starvation during times of dearth is favoured by natural selection (Marlowe and Berbesque 2009). Nutritionally, meat may be a superior foodstuff, but for early hominids the energy required to obtain it precluded its making anything other than an irregular contribution to the diet (Carmody et al. 2011; but see Judson 2017 and Thomas 2017). Observations on carnivory in chimpanzees (Stanford 1998) have led some authorities to propose that hunting by early hominins may have had more to do with status than with nutrition. The diet of the first humans certainly bore no relation to today's meat-rich 'Paleo Diet' fad, and was absolutely not a recipe for nutritional health (Zuk 2013).

Of all the ontogenetic changes during human evolution, the trend toward increased brain size is the most dramatic and far-reaching. This has been studied from every conceivable perspective and it is clear that it has taken a combination of genetic, nutritional, environmental and cultural factors over time to equip *Homo sapiens* with an organ conceptually capable of building the Sultan Ahmed Blue Mosque, composing the Goldberg Variations and flooding the internet with amusing cat videos. Based on life history and metabolic theories, brain evolution can be simulated mathematically in terms of partition of energetic expense between production (learning), maintenance (memory), and execution, of energy-extraction skills (González-Forero et al. 2017). Among the outcomes of this study are models that generate the adult brain and body mass of ancient human scale by what the authors term a 'me-vs-nature' setting, and that relate moderately stressful environments to the development of large brains, reasonably effective skills and metabolically expensive memory. The emphasis on energy here is entirely consistent with the view that a critical driver of brain evolution in hominins is consumption of high glycaemic foods rich in starch (Hardy et al. 2015).

The subject of the genetic and physiological relationship between humans and starch is endlessly fascinating, especially to those of us who need to watch our diet. Of central importance in this liaison is starch structure (Wang et al. 2015). Starch comprises two polymers of glucose: amylose, the $\alpha$-1,4-linear unbranched component, and highly branched amylopectin molecules with both $\alpha$-1,4 and $\alpha$-1,6 linkages. Amylopectin is the more abundant of the two constituents (typically around 75%) and is largely responsible for the semicrystalline structure and lamellar form of native starch granules (Jones et al. 2013). To release the metabolic energy in starch, it is necessary to hydrolyse its component polymers into oligosaccharides and monomeric glucose. The enzyme catalysing this conversion is amylase, and in humans this comes in two forms, one active in saliva (salivary amylase, AMY1) and one secreted into the small intestine (pancreatic amylase, AMY2; Merritt and Karn 1977). The catalytic specificity of both forms of amylase is such that they can cleave α-1,4 linkages between adjacent glucose residues, but not the α-1,6 bonds at the branchpoints in amylopectin. Ingested starch that survives hydrolysis by the amylases passes to the lower gut where it is processed by various α-1,4 and α-1,6 glucosidases. Starch rapidly hydrolysed in saliva and stomach provides a fast glycaemic hit, ramping up blood sugar levels. In diets based on fallback foods, this was and is highly desirable.

There came a time - analysis of sediments in the Wonderwerk Cave complex in South Africa puts the date at around 1 Mya, give or take a few hundred Ky - when *Homo* (probably *erectus*) began using fire (Berna et al. 2012) and early humans started cooking (Wrangham 2009). Heating USOs and other foodstuffs of plant origin enhances their palatability. Exposure to hot water causes starch to gelatinise through a combination of hydration, swelling and collapse of granule structure. Gelatinisation increases starch digestibility in the upper gut and further enhances the release of glucose into the bloodstream. It follows that the factor connecting dietary adaptations, the adoption of cooking, the timing of duplication events at the genetic locus encoding AMY1 and the rapid growth in hominin brain size during the Middle Pleistocene is likely to be the elevated supply of glucose derived from starch (Hardy et al. 2015).

As an aside, I should point out that the organic carbon from which starch is made comes, of course, from photosynthesis. Green tissues export photosynthate as sucrose, the familiar table sugar you stir into your coffee. To make bulk starch from imported sucrose in USOs and other storage organs such as seeds necessitates large quantities of metabolic energy in the form of ATP, complex enzyme machinery and a highly organised intracellular membrane system (Zeeman et al. 2010). These stringent requirements combine to make starch biosynthesis more sensitive than photosynthesis to inhibition by low temperatures and/or water stress. Sucrose, derived from current photosynthesis, that cannot be converted to starch will build up, a physiological condition to which plants needed to adapt as they diversified into cooler environments. Many temperate species have evolved a low temperature-tolerant pathway leading from sucrose towards fructan (Pollock 1986), an alternative storage polysaccharide that deserves a place as an interesting footnote in the starch story. Among the economically important fructan-storing plants are cool-season cereals, pasture grasses, vegetables and ornamentals. In wheat, barley and other temperate grasses, fructans may be accumulated throughout the plant, in foliage, stems and leaf bases, though mature grains contain extremely low quantities and retain the capacity for bulk starch storage. Fructans are polymers based on sucrose to which chains of fructose residues are added. Synthesis of fructans is driven by the mass action effect of a build-up of sucrose concentration and is catalysed by fructosyl transferase enzymes without the requirement for ATP. The long-chain fructose polymer products, which are made and accumulated in the central vacuole of the cell, are water-soluble and therefore osmotically active. They have membrane-stabilising properties and function in temperate grasses, and many other plant species adapted to cold climates, as effective anti-stress metabolites as well as storage compounds. Fructans in the diet are not digestible in the upper gut and are generally metabolised lower down by intestinal bacteria, with sometimes antisocial consequences. In his 1633 revision of Gerard's *Herball*, John Goodyer famously wrote of the fructan-rich USOs of Jerusalem artichoke 'in my judgement, which way soever they be dressed and eaten, they stir up and cause a filthie loathsome stinking wind within the body, thereby causing the belly to be much pained and tormented, and are a meat more fit for swine than man' (Kays and Nottingham 2007).

Generally speaking, there is plenty of evidence to assign special status to the herbaceous plants of grassland origin in the diets of early hominins (Ungar and Sponheimer 2011); but we should acknowledge that trees are known to have been significant food sources for groups of early humans. For example, 1.8 My old remains of date palms, some with trunks that would have been rich in edible starch, have been identified in the Olduvai Gorge site in Tanzania (Albert et al. 2009). The famous Olduvai hominid fossil *Australopithecus boisei* discovered in 1959 by Mary and Louis Leakey and nicknamed 'nutcracker man' was probably a seasonal consumer of savannah tree seeds (Peters 1987). And there are numerous instances known of hazelnuts, pine nuts, acorns, chestnuts and other produce of tree origin amongst the residues in ancient sites as early species of *Homo*, such as *H. erectus* and *H. neanderthalensis*, dispersed during the Pleistocene (eg Shipley and Kindscher 2016). But starch is the thread that runs through the story of the evolutionary origins of the genus *Homo*, and it's grasses that are the starch specialists. As we will see in the next chapter, human evolution was changed utterly when the subterranean starch of USOs took to the air in the form of cereal grains. Grasses bribed humans to join their conspiracy against trees, and starch was the sweetener.

64

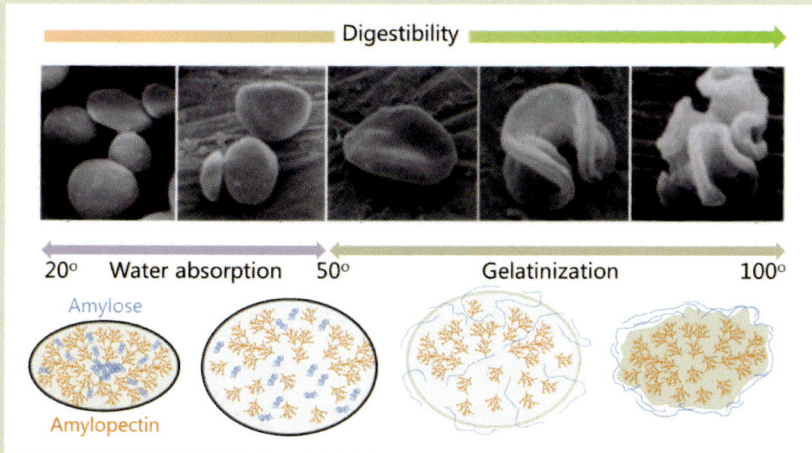

**Accessibility and resistance**

Digestibility

20° Water absorption 50° Gelatinization 100°

Amylose

Amylopectin

Uncooked starch in the raw semi-crystalline form can be hydrolysed by amylases only slowly and incompletely. Cooking renders starch more digestible by disrupting the helical structure of amylose. Denatured amylose molecules are extruded from the swelling starch granule on exposure to hot water, eventually forming an accessible gel surrounding the collapsed remnant of the granule in which the branched amylopectin residues are concentrated (Wang et al. 2015). It was digestion of cooked starch in the upper gut by salivary amylase, which rapidly releases glucose into the bloodstream, that made a critical positive contribution to early human evolution (Hardy et al. 2015). But starch's high glycaemic load has become problematical in the diets of modern developed and developing countries. Digestible starch, together with sucrose and high fructose corn syrup, is a major source of available carbohydrate and implicated in increased risks of dental caries, diabetes, cardiovascular disease and cancer (Brand-Miller 2003). And now here we are (faced with the diseases of affluence and excess, and in complete contrast to our ancestors) in need of starch that passes into the colon largely undigested, thereby promoting better blood glucose control and the growth of beneficial gut microflora (Keenan et al. 2015). For the modern hyperglycaemic descendants of the earliest hominins, resistance is useful.

**Food of the dogs**

Alignment of amino acid sequences of human salivary amylase (AMY1) with pancreatic amylases from human (AMY2) and dog. Over the complete sequence of 511 amino acids, AMY1 and AMY2 are 97% identical, human and dog amylase 90% identical. The rainbow format molecular model is of AMY1, or maybe AMY2 - they have virtually identical 3-D structures.

```
Human salivary    MKLFWLLFTIGFCWAQYSSNTQQGRTSIVHLFEWRWVDIALECERYLAPKGFGGVQVSPP
Human pancreatic  MKFFLLLFTIGFCWAQYSPNTQQGRTSIVHLFEWRWVDIALECERYLAPKGFGGVQVSPP
Dog pancreatic    MKFFLLLSVIGFCWAQYAPNTKPGRTSIVHLFEWRWADIALECERYLAPRGFGGVQISPP
Consensus         **:* ** .*******:.**: ************* ************.******:***

                  NENVAIHNPFRPWWERYQPVSYKLCTRSGNEDEFRNMVTRCNNVGVRIYVDAVINHMCGN
                  NENVAIYNPFRPWWERYQPVSYKLCTRSGNEDEFRNMVTRCNNVGVRIYVDAVINHMCGN
                  NENVVINNPSRPWWERYQPISYKLCTRSGNEDEFKDMVTRCNNVGVYIYVDAVINHMCGN
                  ****.* ** *********:**************.:********** *************

                  AVSAGTSSTCGS  NP SI SF           DFNDGKCKTGSGDIENYNDATQVRDCRLSGL
                  AVSAGTS TCGS FNPG DF PAV  SGW FNDGKCKTGSGDIENYNDATQVRDCRLTGL
                  AVSAGTSS GS  DFPA PP W FNDG   T GSGDIENYNDPYQVRDCRLVGL
                  ********       *        *          ********* ******* .****

                  LDLA   DV R   EYMN     A KH      L  LNSNWFPEG
                  LDLA     MNH ID VA  LDFA WP  IKATI KI  LNSNWFPAG
                  LDLA    LNH  V   KHM DMK   HNLNTRWFPGG
                  ****           *       *   .:  :   ***; *** *

                  SKPFI  DEVI   PP   TGNG  TF K      IRKWN  M  RNWGEGWG
                  SKPFIY VI     FN GAN   R   LKG  GWG
                  SKPFIYQ   LE NW         *  FLKNWG GWG
                  ************  :*:              *.* ** :***:*  ***

                  FMPSDRALVFVDN DN RG    LL T DA LYKMAVGE       VMSSYRWP
                  FVPSDRALVFVDNHDNQRGH            LYKMAVG  I P GFTRVMSSYRWP
                  FMPSDRALVFVDNHDNQRGHGAGGAS   DS YKMGVGF I P GFTRVMSSFRWP
                  *:************* *******          *** ***  *  ***********

                  RYFENGKDVNDWVGPPNDNGVTKEVTINPDTTCGNDWVCEHRWRQIRNMVNFRNVVDGQP
                  RQFQNGNDVNDWVGPPNNNGVIKEVTINPDTTCGNDWVCEHRWRQIRNMVIFRNVVDGQP
                  RYFENGKDVNDWYGPPNNNGIIKEVTINPDTTCGNDWVCEHRWRQIRNMVMFRNVVDGQP
                  * *:**:***** ****.**: ********************************* *****

                  FTNWYDNGSNQVAFGRGNRGFIVFNNDDWTFSLTLQTGLPAGTYCDVISGDKINGNCTGI
                  FTNWYDNGSNQVAFGRGNRGFIVFNNDDWSFSLTLQTGLPAGTYCDVISGDKINGNCTGI
                  FTNWWDNGSNQVAFGRGNKGFIVFNNDDWPLSLTLQTGLPAGTYCDVISGDKIDGNCTGI
                  ****:*************:*.*********   *************************:******

                  KIYVSDDGKAHFSISNSAEDPFIAIHAESKL
                  KIYVSDDGKAHFSISNSAEDPFIAIHAESKL
                  KIYISGDGNAHFSISNSAEDPFIAIHAESKL
                  ***:*.**:**********************
```

Human amylase genes form a cluster on the short arm of Chromosome 1 and comprise pancreatic genes (*AMY2A, AMY2B...*), salivary genes (*AMY1A, AMY1B...*) and a truncated pseudogene *AMYP1* (Dracopoli and Meisler 1990). This region of the genome is a duplication hotspot, and many studies have related copy numbers, particularly of *AMY1*, to the starch content of the diet. Higher *AMY1* copy numbers and protein levels are believed to improve the digestion of starchy foods and buffer against the fitness-reducing effects of intestinal disease (Perry et al. 2007). A remarkable example of increased *AMY* copy number has occurred during dog domestication, which dates to more than 15 Kya. In this case multiple copies of genes for pancreatic amylase (dogs lack *AMY1*) seem to have arisen by parallel adaptation to shared human diets rich in digestible cooked starches (Axelsson et al. 2013).

## Chapter 12 Welcome to the Anthropocene

The Anthropocene is the name given to the most recent stratigraphic age of geological history, in which human activity has become a dominant factor in altering global climate and environment. When they coined the term, Crutzen and Stoermer (2000) associated the beginning of the Anthropocene with the spread of industrialisation in the early nineteenth century. Subsequently, researchers have taken the Anthropocene concept and pushed the onset of the era progressively further back in time. Ruddiman and Thomson (2001) argued for 5 Kya, on the basis of anomalies in the methane cycle observed in ice cores. Taking forest clearance and early agriculture in Eurasia as the Anthropocene starting point dates it to 8 Kya (Ruddiman 2003). For Waters et al. (2016) it's 12 Kya, by which time all of the continents except the South Pacific islands and Antarctica had been settled by humans. If the anthropogenic mass extinction of the Australian megafaunal marsupials in the Pleistocene is taken as the dawn of the Anthropocene, the beginning was as early as 50 Kya (Ruddiman et al. 2015). In the context of the present book, the Anthropocene era could be considered to be congruent with the unfolding tree-grass-human relationship, which would place the onset back even further.

We left species of early *Homo* loading up with high-glycaemic starch from USOs, rapidly growing their brains, learning to use tools and to cook, and beginning the long march out of Africa to colonise the world. In the 200 Ky or so before the widespread adoption of agriculture, *H. sapiens* lived sustainably in loose hunter–gatherer groups without permanent settlements (Gowdy and Krall 2013). The consequence of the transition to agriculture-based societies was the Neolithic Demographic Transition (NDT), an unprecedented population explosion during which the number of humans living on the planet rapidly increased from 4-6 million around 6 Kya to 250 million 4 Ky later (Biraben 2003). Civilisations built on foundations of plant and animal domestication and the husbandry of crops and livestock appeared independently in a few major centres across the world (Vavilov and Dorofeev 1992, Harris and Hillman 2014, Larson and Fuller 2014). In the period 14-9 Kya, the Fertile Crescent, extending over areas of present-day Iran, Iraq, Israel, Lebanon, Turkey and Syria, was the centre of origin for cultivated einkorn and emmer wheats, barley, pasture grasses of the *Festulolium* complex, pea, lentil and vetch; remains of domesticated dogs, goats and sheep are also known from sites in this area. The Yellow and Yangtze River basins were major centres of early agriculture in Asia, with evidence of domesticated rice, millet, rape, hemp, cattle, pigs, dogs and poultry from as early as 11 Kya. Meso-America is the source of another of the world's staples, maize, which was domesticated more than 5 Kya, and North-east Africa was the provenance of sorghum, which dispersed to South-East Asia about 4 Kya. This brief scan of the landscape of early agricultural activity reveals that the same grasses that feed the world today (Warren 2015) were already prominent in the inventory of domesticated plants more than 5 Kya.

The period of transition from hunter-gatherer to agricultural lifestyles was a time of major change in the relationship between humans, grasses and trees. The rise of agriculture, 20 to 10 Kya, coincides with the retreat of the northern ice-sheets at the end of the last great glaciation. Humans found themselves drawn ever further into the skirmishes between grasslands and forests. As the limit of the ice moved north, a tundra grassland-like flora recolonised the land, and herds of grazing animals followed. But domination by grasses was relatively short-lived. The climate continued to warm and the latitudinal tree-line headed northwards (MacDonald et al. 2000). This had the effect of shading out the grasses and driving grazers and tundra further and further to the east and north: and so we find the mammoths making their last stand against climatic attrition and hunting by humans in Arctic Siberia less than 4 Kya (Nogués-Bravo et al. 2008). Post-glacial afforestation was in turn followed by migration of humans, with their farming culture and grazing animals. Thus were the battle-lines drawn for a new phase in the age-old struggle for dominance between trees and grasses.

Plant domestication and the origin of crops has been intensively studied and extensively written about. I intend therefore to be selective, paying particular attention to what happened when the action moved from the subterranean realm of USOs to the aerial world of foliage and seeds. 'All flesh is as grass, and all the glory of man as the flower of grass' the Bible tells us (Peter 1:24-25, KJV). Vegetation phenology is markedly sensitive to latitudinal distance from the equatorial regions where hominins originated (Frenne et al. 2013). A more or less steady supply of USO starch throughout the year becomes more seasonally episodic as the amplitude of cyclical alternations between the vegetative and reproductive phases of plant life-cycles increases. Flowering followed by seed development are visible signifiers of an accessible source of starch. Early humans were aware that starch had taken to the air, and accordingly to have adapted nutritionally and behaviourally, as they began to disperse northwards from Africa into Asia (Finlayson 2005, Nic Eoin 2016). Grass seeds (strictly speaking, they are single-seeded fruits) store starch in a specialised tissue, endosperm. Endosperm differentiation and structure (see Jones et al. 2013) confer properties that make grass seeds particularly suited to supply the dietary needs of a burgeoning human population. In many plants the endosperm is absorbed during embryogenesis and is absent from the mature seed, but in the grasses, endosperm is persistent and represents the major repository of the stored reserves that support seedling growth. Starchy endosperm is unusual in its genotype (it's triploid) and the way it develops, and because its constituent cells are dead at maturity. Grass endosperm is preserved in a dehydrated, mummified state, which makes cereal grains the storable packages of dietary carbohydrate par excellence. Selection for larger seeds with more starch occurred early and rapidly during crop domestication, within perhaps the first 1000 years (Fuller 2007). Thus it was that agriculture provided human civilisation with the fuel for liftoff.

Starch may have been a critical factor in the evolution of *Homo sapiens*; but it is only number two in the list of bulk sources of glucose. Half of global biomass is accounted for by cellulose, the polysaccharide framework of plant cell walls. Cellulose consists of β-1,4 linked glucose monomers (Jones et al. 2013). Humans lack digestive enzymes that will cut this chemical bond and consequently cannot directly use this source of sugar; but grasslands, abundant sources of cellulosic vegetation, are full of grazers and browsers that can. In this regard, we may regard the wild animals and domesticated livestock exploited for meat and milk by hunting and herding, as intermediaries employed by humans to access the bioenergy locked up in the cellulose of grass leaves. Of course, it's usual to think of meat and dairy products in the modern diet primarily as sources of protein and fat. But in the hierarchy of nutritional priorities, sufficient calories comes first, and under conditions of subsistence or inanition, when the supply of metabolisable sugars is inadequate or marginal, the carbon skeletons of protein amino acids will be used for bioenergy generation after stripping off the nitrogen in the form of ammonia (Thomas 1988). It follows that the first settled environments of hunters, trappers, early herders and mixed farmers would tend to have been enriched in nitrogen released by respiration from dietary animal protein. The fertile, disturbed habitats surrounding early settlements and in cultivated fields were an open invitation to synanthropic plants, invading grass crop progenitors with the advantageous traits that we now recognise as the basis of the 'domestication syndrome' (Abbo et al. 2005, Cunniff et al. 2014). Moving from a starch to a starch+cellulose economy would have represented a prodigious increase in the capture of solar energy via photosynthesis and it is reasonable to suppose that this was an important driver of the NDT.

Civilisation is built on agriculture, and agriculture is built on grass. What, then, of trees and the development of early societies? Wood is about 40% cellulose, but lignification renders its glucose nutritionally inaccessible to all but the most specialised insects, fungi and microorganisms. To release wood's store of energy, we have to resort to burning it. We are familiar with stories in the news media, literature and motion pictures of previously undiscovered tribes living in remote forest areas (usually deep in the jungles of South America, Africa or the Far East). At the source of this appetite for tales of lost civilisations is a fascination and even obsession with the idea of a prelapsarian world beyond the reach of corrupt modernity. According to the Brazilian National Indian Foundation (FUNAI), Brazil's native peoples consist of around 400,000 individuals (0.2% of the total population) living in disconnected areas across 12.5% of the national territory. About 70 isolated indigenous groups have been identified in recent years (Shelton et al. 2013). A typical community in the western Amazonian state of Acre, leading what was described as a Neolithic way of life, hit the news after José Carlos Meirelles of FUNAI announced contact in 2008. Plots of maize, manioc and bananas observed surrounding the cluster of communal huts have been estimated to be capable of contributing to the sustenance of a community of up to a hundred. As a general rule, a small social grouping can make a living from hunting, pannage, 'wood-pasture' and consumption of forest fruits and nuts but, even at this limited scale, sustainability seems to require a degree of tree clearance to make way for crop cultivation. New archaeological techniques, particularly LiDAR (LIght Detection And Ranging), are discovering ancient forest civilisations, often described as lost jungle cities (Horton 2016). Although these extensive settlements appear to be isolated and self-contained, it is becoming clear that they conform to the model of social and cultural development and function established for the centres of archetypal agricultural civilisation. Ruins of the 2 Ky old Mayan city of El Perú-*Waka'* in the rainforest of northwestern Guatemala have been archaeologically studied since around 2003 (Eppich 2015). La Ciudad del Jaguar, a prehistoric city buried beneath tropical vegetation, flourished in a densely populated region of the Mosquitia territory, Honduras (Fisher et al. 2016). Both cases are typical of 'lost cities' in deep forest locations which, rigorous archaeological analysis reveals, were in fact situated in extensively managed landscapes and dependent for their expansion and culture on agricultural production in the environments that surrounded them. In this respect, the city's supply of food was plainly organised on the classic Thünen principle (von Thünen 1966, Lloyd and Dicken 1972), in which the 'isolated state' would be the focus of agricultural activity organised in a series of rings. As in the Ancient Mariner's tale of 'water, water everywhere, nor any drop to drink', forest-dwelling hunter-gatherers are surrounded by huge reserves of inaccessible lignocellulose that circumscribe their world. Such communities are severely constrained in size and cannot achieve the population densities and cultural expansion unleashed by the rise of agriculture (Ortman et al. 2015).

In the Hitchcockian drama that is the Anthropocene, starch and cellulose are the MacGuffin. They are the means by which agriculture meets the basic necessities of humankind - the physiological foundation in Maslow's (1943) hierarchy of needs. A consequence of freedom from the unrelenting demands of subsistence is the rise of cultural engagement with higher-order desires in the Maslow hierarchy, those concerned with esteem, self-actualisation and self-transcendence: a change of focus from, in the words of Vladimir Nabokov, the personal and physical to the universal and metaphysical. The next chapter considers what happened when ritual, religious and creative expression entered into the already conflicted relationship between people, trees and grasses.

**Grazing animals turn photons into food**

The primary purpose of agriculture is to convert solar energy into the carbon compounds that feed and power humanity (Loomis and Williams 1963, Amthor 2010; fossil fuels are simply yesterday's photons). Humans do not have direct dietary access to cellulose, the most abundant plant product of light and carbon capture, but hind-gut fermenters and ruminant grazers do. Humans use animals as intermediaries to harvest the solar energy locked up in cellulose (Thomas 2017). Our dependency on grasses is thus two-fold: directly through the starch in cereal grains, and via animal proxies that convert cellulosic forage into meat, milk, wool and hide. As Thomas Browne expressed it, we are 'grass carnified' (Bennett 2010). In modern intensive livestock production, one thing grazing animals do not do is graze; instead they feed on grain and other produce from remote cropping systems, clearly a practice of dubious environmental and economic sustainability (Duru and Therond 2015). Demands for an end to meat and dairy consumption on hard-line environmentalist and ethical grounds, without any allowance made for ecology and culture, don't make much sense either (Stenfeld et al. 2005, Hsaio 2015). In the end, it comes down to how global humanity gets its photons. Only starch and cellulose together can meet this need on anything like a sufficient scale.

## Chapter 13 Fear and madness in and out of the forest

I arrived at this chapter ready to re-tell stories our distant ancestors have passed to us, but I was quite unprepared for the difficulty of seeing the wood for the trees that researching the literature reveals. A scientist like me should heed Dante when venturing into this 'savage forest, dense and difficult', and enlist at least one Virgil for guidance. My initial perspective on the metaphorical and metaphysical roots of the tree-grass-human nexus in early societies and cultures is delineated in Archer et al. (2014), a study in which Jayne Archer, Richard Marggraf Turley and I placed the history of The Field at the centre of the physical and imaginative relationship between people, food and landscape. For Harrison (1992), it's the interface between forest and city that is definitive in the development of human culture and creativity. Schama's (1995) prodigious meditation on the meaning of landscape in Western (it's overwhelmingly Western) civilisation divides the world into wood (he is marvellously informative on forests), water and rock. That venerable anthropological institution, *The Golden Bough* (Frazer 1922), takes a different direction, freely moving between trees and corn as mystic and sacred tokens of humankind's progression from magic to religion to science. And then there's *The White Goddess*, Robert Graves's 'historical grammar of poetic myth', in which the title of the second chapter echoes that of the present book: 'The battle of the trees' (Graves 1948). On the whole, the extreme richness and intrinsically paradoxical nature of the canon makes it a most un-Virgilian guide. The only practical way forward I can see is to pick out themes, à la carte-style, that illustrate the central proposition of the present book, namely the exploitative, often even downright aggressive, interplay of people, grasses and trees.

Someone once said that if lightning is divine wrath, trees must have done something extremely bad to make the Gods so very angry. On the other hand, the earliest pages of the Old Testament tell us that God loves a tree more than Man. It seems that, after the incident with the apple and the serpent, God wants humans out of the forest of paradise to where He can keep an eye on them. In the words of Giambattista Vico (1668-1744), whose archetypal writings on trees are much quoted by Harrison (1992), 'the forests become monstrous, for they hide the prospect of God'. John Evelyn (1664) imagined the estrangement between trees and people in the pre-biblical era thus: 'Acorns, *esculus ab esca* (before the use of wheat-corn was found out) were heretofore the food of men, nay of Jupiter himself, (as well as other productions of the earth) till their luxurious palats were debauched'. Expulsion from Eden is followed by the second Fall of Man and the Curse of Cain under which, as Genesis 3 and 4 (KJV) tells us, God condemns humans to enslavement by the grasses that feed us and our animals:

> '...thou shalt eate the herbe of the field. In the sweate of thy face shalt thou eate bread, till thou returne vnto the ground ... Abel was a keeper of sheep, but Cain was a tiller of the ground. And in the processe of time it came to passe, that Cain brought of the fruite of the ground, an offering vnto the LORD. And Abel, he also brought of the firstlings [firstborn] of his flocke, and of the fat thereof: and the LORD had respect vnto Abel, and to his offering. But vnto Cain, and to his offring he had not respect: and Cain was very wroth, and his countenance fell.'

Not content with trafficking us to the grasses, God, it seems, proceeded to favour pastoral over arable farming. Maybe He prefers cellulose to starch, who can tell what goes on in the mind of deity? Tudge (2004) has pointed out that, 'it was the shepherds, successors of Abel, who attended the birth of Jesus. No one turned up with a sack of barley'.

The Fall of Man not only began with human estrangement from trees; and not only did it sow murderous discord between the livestock farmer and his arable brother; but it also turned agricultural weeds loose on the world as the incarnation of Original Sin (Edwards 1789, Drury 1992, Archer et al. 2014). In this case, Christian tradition can be traced back to classical accounts of the loss of a Golden Age as early as Hesiod's *Works and Days* (c. 700 BCE). In Book 1 of the *Georgics* of Virgil (c. 29 BCE), the emergence of weeds is taken to be a sign of translation of Astraea, goddess of justice, from earth to heaven. Henceforth, it is explained, farmers will need to toil at ploughing the soil and removing weeds from crops in order to keep hunger at bay (Thomas et al. 2016). Weeds, species that went along for the ride during crop domestication, throw a revealing light on the nature of the cultural relationship between plants and people. One such fellow-traveller is the poisonous cereal analogue darnel (*Lolium temulentum*). Abundant literary, historical, religious, medical, and scientific sources show it to be an anthropophytic annual grass weed that evolved from a perennial ryegrass progenitor and was subject to the same human-mediated selection pressures as the ancestral cereal species it infested (Thomas et al. 2016). Ryegrasses (*Lolium* spp.), and their close relatives the fescues, are major pasture grasses of northern temperate regions. Their centre of origin was the same area of the Middle East as wheat and barley, whence they spread alongside the diversification of arable and livestock farming (Thomas et al. 2011). The invasion of cereal fields by the descendant of a perennial pasture species introduces grass-on-grass internecine conflict - the Curse of Cain in vegetable form - as a new layer of complexity to the war on trees.

The toxicity of darnel grains (the Mark of Cain perhaps?) is due to a cocktail of phytochemicals secreted by genetically complex endophytic fungi closely related to ergot (as described in Chapter 9). Darnel's reputation as a poisonous cereal look-alike that corrupts the food-chain made the plant a malign symbol of religious heterodoxy and sedition. An enduring metaphor is the biblical parable of wheat and tares (Matthew 13: 24-30; tares is one of the many alternative names for darnel). Because darnel/tares mimics the cereals it infests, the name of the plant has been invoked to denounce subversive religious dissent, but also by self-proclaimed heretics as well as irenic voices calling for tolerance. Thomas et al. (2016) explain how the name 'tares' came to be substituted for 'darnel' at the time of the second Wyclif bible (c. 1395), and how such a deliberately ambiguous nomenclature was associated with the rise of lollardy (the 14th century heretical Catholic reform movement whose name is suggested by some accounts to be derived from *Lolium*).

In scriptural exegesis, husbandry manuals, and literature in the georgic tradition, darnel symbolises The Fall of Man. For William Shakespeare and his coevals in the Early Modern period, darnel came to speak of the fall of *a man*, becoming a potent metaphor for treachery and psychological disintegration projected onto a world perceived to be in disarray. Of several works by Shakespeare in which darnel makes its presence felt, *King Lear* is particularly pertinent to the subject of the present chapter (Archer et al. 2012). The display of darnel in the mad king's crown of 'idle weeds' makes an acute psychological and political point about doctrine of the king's two bodies, intertwining the body politic and the body natural ('l'Etat, c'est moi'). Allusion to darnel declares that a deranged king and a sick land have become one. In the tragedy of *Lear* we are reminded of one of the most potent biblical stories concerning trees, grass and kingship. Chapter 4 of the Book of Daniel tells of King Nebuchadnezzar's dream of a tree:

'A tree in the midst of the earth,
And its height was great.
The tree grew and became strong;
Its height reached to the heavens,
And it could be seen to the ends of all the earth…'

Then in the dream, the passage continues, 'a holy one, coming down from heaven' commanded:

'Chop down the tree and cut off its branches…
Nevertheless leave the stump and roots in the earth,
Bound with a band of iron and bronze,
In the tender grass of the field.
Let it be wet with the dew of heaven,
And let him graze with the beasts
On the grass of the earth.
Let his heart be changed from that of a man,
Let him be given the heart of a beast...'

In Grigori Kozinsev's celebrated 1971 Russian language film of *King Lear* (*Korol Lir*), the mad King is shown crawling through wheatfields like a grazing animal (Marggraf Turley et al. 2010; it is striking that the symptoms of Lear's madness are those of someone who has eaten darnel). A year after his dream, Nebuchadnezzar in his insanity was cast out into the forest. Shakespeare has plenty to say about the troubled relations between humans and trees. Several of the plays are set in woods, which are often places of, at best, enchantment and at, worst, extreme peril. Perhaps the harshest words come from one of the most violent of the tragedies, *Titus Andronicus*. Here the woods are described as 'ruthless', 'dreadful', 'gloomy' and, most shocking of all, 'made for murders and rapes'. One is reminded of the Evil Forest 'alive with sinister powers of darkness' in Chinua Achebe's 1958 novel of Nigerian village life *Things Fall Apart*. When it comes to encounters with the wildwood, it seems that fear and madness are truly universal atavistic human attributes.

Before the imposition of the feudal system in medieval England there were common rights of access to land, including woodland. In the period from the Norman invasion to the 13th century, Forest Law was applied to a large area across the country. Forest was reserved for the 'royal pleasure', which generally meant hunting. Not only was the pursuit of game forbidden, but even cutting wood, collecting fallen timber or harvesting berries were illegal acts, punishable in the harshest manner. Steffes (2016) makes the connection between Forest Law and the topography of King Lear's insanity, a monarch alone in a blighted wilderness. The deep etymology of 'forest' connects it with words such as 'foreign' and 'forfeit' and designates a place outside or beyond the reach of common law (Harrison 1992). The importance of *la chasse* in French rural life is rooted in a history of social protest reaching back to the Revolution. It asserts an individualist ideology in opposition to the influence of the bourgeoisie, particularly the urban classes during the 19th century, and even today is driven by cultural memories of the French equivalent of forest law and tensions over the enclosure of wilderness for agricultural production (Mischi 2013). Thus not only was the forest historically a symbol of sin and torment, it was, literally and legally, a no-go area.

According to Dr Jianchu Xu of the World Agroforestry Center, 42-51% of global tree cover has been lost over the course of human history ($2.4$-$3.4 \times 10^9$ ha). The present-day area of cropland is estimated to be of the same order, around $5 \times 10^9$ ha (Goldewijk et al. 2011). It looks superficially as if there must be a rather direct cause-and-effect relationship here, with humans recapitulating the evolutionary era of forest burning, thereby replacing trees with man-made grasslands. Over the course of human history, crops and pastures have certainly been established on a vast scale at the expense of woodland. But according to some archaeologists, too much emphasis is placed on agriculture as a cause of deforestation. In the words of Bradley (2005) 'archaeology is impoverished unless it sheds its fixation with food production'. As Harrison (1992) discusses at length, the eponymous hero of the *Epic of Gilgamesh*, the earliest known work of literature (dated from 1800 to 1000 BCE) slays the monstrous demi-god Humbaba and proceeds to fell the trees of Cedar Mountain, of which the creature was guardian. This act of deforestation has nothing to do with clearance for agriculture and everything to do with fear of mortality and the lust for fame. The *Thebaid* (a Latin poem written by Publius Papinius Statius around 80-92 CE) retells the ancient tale of Seven against Thebes, with its origins in Bronze Age culture a generation earlier than the Trojan Wars. Books V and VI recount the killing of the infant Opheltes by a snake and describe in great detail the funeral pyres built by the Argive Seven for the child and the slain serpent. Statius identifies the source of the wood for the pyres as an ancient sacred grove and conceives this act of deforestation as a deliberate deed of sacrilege and an expression of the excesses of war (Ganiban 2013). Similar episodes of tree-felling with the object of desecration occur in works by Ovid and Lucan. Notoriously, strategic defoliation and deforestation were employed to devastating effect in the 1955-1975 Vietnam War (Lang 2001). Up to the present day, there is widespread deforestation, particularly in Southeast Asia, to provide fuel for the pyres of traditional (particularly Hindu) funerary rituals. Around 7 million bodies are burned every year in India and Nepal, each requiring an average of 550 kg of wood. As well as its effect on forest biomass, outdoor cremation is causing concern as a source of air pollution and emissions with significant climate-forcing influence (Chakrabarty et al. 2013).

In conclusion, for much of human history, the wildwood has been seen not only as an obstacle to settlement and agriculture but as a space at once sacred and the embodiment of evil (Archer et al. 2014). The Gnostic and Manichean belief in the goodness of a trimmed and orderly man-made environment, in contrast to untamed Nature as the epitome of devilry and paganism, is a recurrent theme in folklore and religion. Celtic mythology tells of the 'Green Man' or 'Wodewose', famously rendered into literature in the late-14th-century Middle English poem *Sir Gawain and the Green Knight*. In the tales of Red Riding Hood, Hansel and Gretel, Sleeping Beauty and so on, the terrors of the forest feature throughout the work of Charles Perrault and the Brothers Grimm (Harrison 1992, Schama 1995). Tolkien has some particularly fine descriptions of the perils of ancient woodlands (Saguaro and Thacker 2013). The Jungian fear of the wildwood persists even in the modern era, as we see in the madness of Kurtz in *Heart of Darkness* (Conrad 1899), Mr Todd in *A Handful of Dust* (Waugh 1934) and Percy Fawcett in *The Lost City of Z* (Grann 2009). The forest as a place of dread and folly is the subject of several theatre pieces and films, including *Fitzcarraldo*, *The Blair Witch Project*, *Deliverance* and *Into the Woods*. In this connection, as a child I always found the song *The Teddy Bears' Picnic* unsettling because of its strangely sinister minor-key melody and the warning of dire consequences 'if you go down to the woods today'.

**The Green Man**

Green Man vomiting trees, Kilpeck Church, Herefordshire (12th century)

Landscape with anthropomorphic tree (Pietro Ciafferi, 1600-54).

Nebuchadnezzar as a Wild Man (William Blake, 1795) – Tate Britain

Green Man Festival Brecon, Wales

The dangers of the forest are personified in the traditional figure of the Green Man, or Wild Man of the Woods, who recurs in a diversity of forms in literature, art, culture and ritual (Frazer 1993). He is frequently represented in religious spaces. The 12th century sculpture of the Green Man on the capital of the south door of the Church of St Mary and St David, Kilpeck, is in the typical form of a grotesque head with vegetation spewing from its mouth. The figure about to spoil someone's picnic was created by Pietro Ciafferi (also known as Lo Smargiasso – 'The Swashbuckler'), a Tuscan painter better known for his seascapes. William Blake, who famously represented the mad king Nebuchadnezzar as a Wild Man, was the poet of the 'forests of the night', within which the fearfully symmetrical tyger burned bright, and the illustrator of Dante's 'cammino alto e silvestro' (the deep and woody way to Hell; Tambling 2005). Even today variations on the theme of the Green Man are the subjects of observance and ritual around the world. The annual Brecon Beacons Green Man festival is a celebration of music and arts with an independent, non-corporate ethos.

## Chapter 14 A reconciliation, of sorts

There were groves. There were coppices. There was kindling. There was timber. There was a little food. There was sport. But for much of human history, there was Nature, and there was Culture. Not only was woodland not nurtured, and hardly considered as a resource to be husbanded, it was a fearsome place, the embodiment of Augustinian sin and malevolent dreams. Nevertheless, from early times, a few tree species entered into domestic arrangements with people. In the Mediterranean region, olive and edible fig were cultivated as early as 4000 BCE and myrrh from around 1500 BCE. Orchards in China date from 2000 BCE. There is evidence of some exchange of germplasm (of walnut, almond and sweet chestnut, for example) between the eastern Mediterranean and Asia dating back to a few centuries BCE (Turnbull 2009). It is in the Early Modern and Renaissance era, however, that the great expansion in silviculture begins, when 'Europeans came to look at, engage with, and even transform nature and the environment in new ways, as they studied natural objects, painted landscapes, drew maps, built canals, cut down forests, and transferred species from one continent to another' (Cooper 2014). Deforestation leading to shortages of timber and forest products was an early stimulus to tree cultivation. Toynbee (1976), considering the history of Western Christendom in the period 634-756 CE, states that the needs of shipbuilding and architecture, and the demand for fuel to heat baths, greatly depleted woodlands in the Mediterranean region at this time, denuding hills and mountains and reducing the area suitable for agriculture.

Wiersum (1997) proposed a model for the phases of domestication that lead from natural forest to tree crops. The protocols by which woody species of tree-dominated ecosystems become domesticated are only partially analogous to those applicable to field crops derived from open grassland environments and pioneer vegetation. They include a trend from 'wild' to managed forests, subsequent enrichment with exotic species and ultimately the establishment of monospecific plantations. Wooded environments often have their own unique attributes that demand special treatment: for example, sacred groves, resource-enriched natural forests and centres of biodiversity. Appreciation of the special ecological and cultural significance of forests has impelled the rise of movements such as Pro Silva Europe (http://www.prosilva.org/ [accessed 4 October 2017]), Positive Impact Forestry (MacEvoy 2004) and Conservation Arboriculture (Dujesiefken et al. 2016).

The century or so running up to the French Revolution in 1789 is often referred to as the Enlightenment, or Age of Reason in Europe (O'Hara 2010). Its influence extends well beyond this timeline. Harrison (1992) calls it the 'post-Christian era'. As Porter (2001) remarks 'We are still trying to solve the problems of the modern, urban industrial society to which the Enlightenment was midwife, ... largely [drawing] upon the techniques of social analysis, the humanistic values, and the scientific expertise which the philosophes generated. We remain today the Enlightenment's children.' From the perspective of the uneasy, even fearful, history of human engagement with untamed nature, it was the time when *Homo sapiens* at last began to sleep with the lights off. The Enlightenment was a period of political, agrarian and technological revolution. It was defined by the introduction of new agricultural methods and crops, and the enclosure of hitherto communally-owned land. The profound socioeconomic consequences have been subjects of a large literature and I don't intend to consider them in detail except in so far as they can be shown to have directly disturbed the relationships of people, grass and trees.

Crop rotation and the use of root crops and clover alongside cereals began to be practised in the 16th century but took off in a big way from around 1670 (Overton 1985): this, together with reclamation, drainage and the near doubling of the area of land under cultivation, resulted in a several-fold increase in cereal yields over the subsequent two centuries. Enclosure had been underway in England since the 15th century, but accelerated from the mid-18th so that common ownership of land had been largely abolished across the greater part of the middle of the country by 1820 (Gibson 2010). Landowners justified the process in the name of agricultural improvement, but the fact is that international demand for English wool made sheep more profitable than crops, particularly as livestock farming is much less labour-intensive than crop agriculture. The Highland Clearances are a brutal example of population displacement from rural areas to maximise economic return (Richards 2000). It was a tragic consequence of the revolutions in agriculture and land ownership that ultimately the price of corn soared and this, together with the cost of the Napoleonic wars, resulted in famine. The flames of agrarian violence were fanned, not only in England and Scotland but also in Ireland, where enclosure had resulted in Anglo-Irish landlords holding 95% of all land by the early 19th century (Christianson 1972, Powell 2005). Even Wales had its own moment of protest, in the shape of the Rebecca Riots (1839-1844; Molloy 1985).

In a sense, the tension between the pastoral and the georgic continues the age-old theme of factional civil war within the grasses. It's reflected in the language of the period of rural transition: we see it in Cobbett's (1833) insistence that agriculture meant exclusively the cultivation of crops, whereas practitioners of livestock husbandry were not farmers but 'graziers'. John Cowper, in his 1732 essay 'Enclosing Commons', wrote 'who among country people live lazier lives than the grazier and the dairyman? All the dairyman has to do is to call his cows together to be milked!' In considering the literary and socioeconomic history of enclosure, we have argued (Archer et al. 2014) that the assault on common ownership has carried on into the present century, pursuing people as they have moved en masse from countryside to city, and kettling them in ever-diminishing public spaces while zones of exclusion and surveillance grow inexorably. But that's another story.

The Age of Discovery in Europe spans the period of the Enlightenment, running from the Renaissance to the Industrial Revolution. Voyages of exploration literally widened human horizons and, over about 300 years, the course to full globalisation was set. The factors contributing to the burgeoning of human experience and ambition were multi-faceted and complex, comprising inter alia: piracy, both private enterprise and state-sponsored; the urge for scientific discovery; the growth of mercantilism and its refutation by the likes of Adam Smith in favour of free trade and capitalism; the rise of imperialism; and the Great Game between European powers, and latterly America, as nations competed for territory, resources and political influence (Harley 2004). Here is hardly the place for a dilettante scientist to wade deeply into these waters, which are choppy, treacherous and the haunt of shoals of ardent professional economic historians. Instead I'll address three enduring implications for the relationship between people and plants, namely: the demand for timber; the expansion of interest in gardens and aestheticised landscapes; and the Romantic redefinition of affinities between humans and the natural environment.

Voyages mean ships and, even some time after the Industrial Revolution had introduced new materials based on iron, shipbuilding was absolutely dependent on supplies of timber. Since international relations consisted of sending fleets of wooden vessels to every part of the globe, such was the demand for trees that governments during the Age of Discovery were in constant fear of 'timber famine' (Warde 2006). Whether Europe was ever close to a deforestation crisis before the Napoleonic Wars is disputable, but the perception was enough to invoke political action. A momentous event that influenced the attitudes of people (particularly those in High Places) to trees was publication of John Evelyn's (1664) *Sylva*. Evelyn's Diary notes that his *Discourse concerning Forest Trees*, written 'upon occasion of certain queries sent to us by the Commissioners of his Majesties Navy', was delivered to the Royal Society on 5 October 1662 and, along with Robert Hooke's *Micrographia*, was one of the Society's first two published books. According to Nisbet's (1908) introduction to the 4th edition of *Sylva*, between 1603 and 1660 the tonnage of the English fleet surged from 17,110 to 57,463, raising fears that timber stocks in naval dockyards were becoming exhausted. In the century or so from 1608 the supply of useable timber from the New Forest and other sources fell by 90%. The *Sylva* is 'gracefully written in nervous English and in cultured style, ornately embellished according to the then prevailing custom by apt quotations from Latin poets, it contains an enormous amount of information in the shape of legends and of facts ascertained by travel, of observation, and of experience'. It recognises that trees are a valuable national resource and, importantly in a mercantile age, that woodland is an investment that, if carefully managed, delivers assured returns. Nisbet (1908) quotes from Walter Scott's *Heart of Midlothian* (1818), where the Laird o' Dumbiedykes tells his son 'Jock, when ye hae naething else to do, ye may be aye sticking in a tree; it will be growing, Jock, while ye're sleeping'. The *Sylva* systematically describes the botanical characteristics, propagation and cultivation of some 60 tree species. In pride of place, as would be expected of a work inspired by the needs of the Navy, is the oak.

In their wide-ranging study of biological exploitation, Hatcher and Battey (2011) have discussed in detail the evolution, physiology and many uses of the oak, which they, in the tradition of John Evelyn and Edmund Burke, identify as a cultural icon. They cast a sceptical eye over the whole subject of the demand for and supply of timber for shipbuilding and the rhetoric of Hearts of Oak in the

English national character. From ancient times, across the geographical range of *Quercus* spp., acorns have been valuable food sources for humans and their animals, and for wild woodland fauna. The European tradition of pannage, fattening pigs by driving them into forests to forage for tree seeds, has been practiced since prehistory (Parsons 1962, Hamilton et al. 2009). The particular durability of oak timber is accounted for to a great degree by the extent to which the tree accumulates tannins. Tannins are phenolic compounds, chemically related to the anthocyanins and flavonoid pigments of many fruits and flowers and to the lignin polymers that confer mechanical strength on the walls of wood cells (Jones et al. 2013). Because they bristle with charged chemical groups, tannin molecules avidly bind, coagulate and denature proteins. This property underlies their function in defending the plant from attack by pests and diseases, and also accounts for their astringency when consumed in bitter-tasting food and drinks such as cranberries, tea and oaked red wine. Oak bark is particularly rich in tannins and has traditionally been used on a large scale for (the clue is in the name) tanning animal hides to make leather, an industry that ranked with shipbuilding, metal crafts and wool production in 16th and 17th century England (Clarkson 1960, Hatcher and Battey 2011). Another oak whose bark has traditionally been economically important in the Iberian peninsula since Moorish times is *Quercus suber*, the cork oak (Parsons 1962).

As long-haul sea travel gathered pace from the early modern period, so too did the quest for new plants and the introduction of exotics to new environments. The garden historian Maggie Campbell-Culver (2004) has recorded an explosion in the numbers of ornamental species landing in England during the 17th and 18th centuries, the spoils of plant-hunting expeditions sponsored by the insatiable expansionism of European nations. Among the tree species she documents are *Abies* spp. (firs), *Aesculus hippocastanum* (horse chestnut), *Larix decidua* (European larch), *Morus rubra* (red mulberry), *Rhus* spp. (sumacs), *Cedrus libani* (cedar of Lebanon), *Robinia pseudoacacia* (black locust), *Platanus occidentalis* (American sycamore), *Taxodium distichum* (swamp cypress), *Citrus* spp. (lime, lemon, citron, sweet orange, bergamot, pomelo, tangerine), *Prunus lusitanica* (Portuguese laurel), *Liriodendron tulipifera* (tulip tree), *Rhamnus alaternus* (Italian buckthorn), *Juglans nigra* (walnut), *Platanus x hispanica* (London plane), *Cupressus lusitanica* (cedar of Goa), *Cornus* spp. (dogwoods), *Acer* spp. (maples), *Coffea arabica* (coffee), *Quercus* spp. (oaks), *Populus* spp. (poplars), *Picea maritima* (black spruce), *Magnolia* spp. (magnolias), *Ginkgo biloba* (ginkgo), *Betula* spp. (birches), *Tsuga canadensis* (Eastern hemlock), *Pinus* spp. (pines), *Fagus sylvatica* f. *purpurea* (copper beech), *Pistacia vera* (pistachio), *Leptospermum* spp. (tea trees), *Araucaria* spp. (monkeypuzzles), and *Diospyros kaki* (persimmon); not to mention a multitude of shrub species. One can easily imagine that this enormous influx of non-natives and exotics would have revolutionised ideas about the place of trees in the rapidly changing environment of the post-Christian period in Europe.

At the same time, tree species yielding tradable commodities were spreading across the world from their native environments, generating immense wealth for plantation owners and merchants, but often at the expense of severe social and environmental damage arising from imposed changes in land-use and oppression of indigenous people (Collingham 2017). Examples include rubber from Meso-America (Tully 2011), oil palm from West Africa (Berger and Martin 2000), cocoa from South America (Young 1994, Warren 2015) and coffee from Ethiopia (Pendergrast 2001, Warren 2015).

Incidentally (stand back, axe being ground), isn't it strange that the vast scale on which endemic plants have been moved around the planet since the dawn of the Age of Exploration seems to be of such little concern? This despite the fact evolution has equipped these species with whole genomes that made them successful competitors and survivors in their native habitats, and (as we know only too well from invasives - Weber 2003) quite liable to become aggressive and harmful in their new homes. Yet the mere suggestion of crop improvement by a precise and well-intentioned change in a gene or two by geneticists and plant breeders causes conniptions among those of a 'green' persuasion, many of whom will happily fill their gardens with alien genomes from the other side of the world. Lost for words. Now excuse me while I spend a little time with my piano to re-establish the Zen-like calm necessary to carry on with this chapter.

**|| |||**

To continue: it seems, therefore, that by the time we get to the 18th century, the atavistic fears of the forest, and the slash-and-burn legacy of siding with the grasses, began to soften. The Enlightenment, the rise of Romanticism, and appreciation of the Sublime, combined to effect a kind of reconciliation between humans and trees. Explorers like Alexander von Humboldt, who was captivated by his experience of the Hylea Amazonica (Hadley 2005), began to regard forests not as hellish places to be tamed and exploited, but as natural wonders in their own right (Helferich 2004). The pastoral movement saw the rehabilitation of trees, which became aesthetic objects, taking their place in the classical English-style landscapes created by the likes of William Kent, William Shenstone and Capability Brown and in paintings by Thomas Gainsborough and others (Ruff 2015). Trees became benign decorative features, enhancing the informality of the scene and contributing to re-creations of the Italian landscape paintings in fashion at the time. More to the point, it's significant that the plants we see in these idyllic prospects are perennials. The harmony and serenity of these landscapes represent a truce of sorts between grass and tree. The trees are, of course, almost timeless – in fact, the oak alongside Mr and Mrs Andrews in Gainsborough's celebrated painting of the young landowners, though no longer viable, is still standing today on the Auberies Estate Farm in Essex (Primary Teachers' Notes 2014-15). The pasture grasses are clonal perennial communities with lifespans that might match or exceed those of the trees with which they share the landscape (Thomas 2013).

A striking feature of the pastoral world is that no-one seems to do any work. People lounge around and generally present an air of privileged idleness. They can do this because perennial plants don't need to be tended, cultivated and harvested – they look after themselves, effortlessly passing through a seemingly unending cycle of renewal with the ebb and flow of the seasons. As we saw in Chapter 3, trees can control their own architecture with little intervention from humans. And pasture grasses are kept in trim by grazing animals, which are effectively solar-powered self-drive lawnmowers. This releases shepherds and shepherdesses to spend their time striking aesthetic poses and dallying amorously. Alexander Pope's influential essay 'A discourse on Pastoral Poetry' made the connection with the classical era:

> 'The original of Poetry is ascribed to that age which succeeded the creation of the world: And as the keeping of flocks seems to have been the first employment of mankind, the most ancient sort of poetry was probably pastoral. It is natural to imagine, that the leisure of those ancient shepherds requiring some diversion, none was so proper to that

solitary life as singing; and that in their songs they took occasion to celebrate their own felicity...We must therefore use some illusion to render a Pastoral delightful'.

Pope was writing in 1704 about poetry and a mythical Golden Age, but he captured contemporary sentiment, in which entire landscapes could be re-imagined as Arcadian dreamworlds.

Alongside the idealised mise en scène of the Arcadian idyll, however, there was, and always has been, the harsh reality of the georgic life with arable agriculture at its centre. The pastoral mood is – at least superficially – peaceful, whereas that of the georgic is one of barely suppressed violence. If pastoral celebrates human and nature in harmony, the georgic sees that relationship as one of ongoing hostility and struggle, as natural forces frustrate our ability to bend the land to our will. Such are the conjoined destinies of crops and humankind that the cereals which feed us must be planted, tended, protected from pests and diseases, and harvested anew every year, making for a remorseless calendar of sweat and toil, even in the present era of industrialised farming (Archer et al. 2014).

**The oak as Romantic and political symbol**

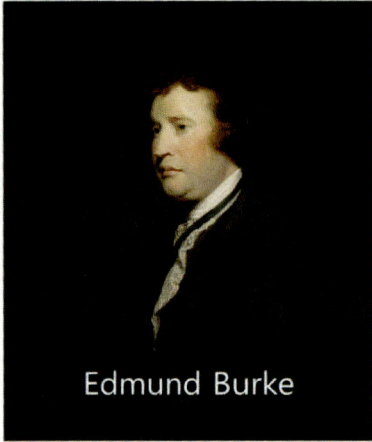

Edmund Burke

The logo of the Conservative
Party of Great Britain

Created by and for elite city dwellers, the English pastoral is intrinsically conservative and reactionary. Edmund Burke, in his *Reflections on the Revolution in France* (1790), used that enduring emblem of Englishness, the oak tree, to naturalise his opposition to the politics of revolution. He argued that the English constitution should be like a slow-growing oak tree that 'moves on through the varied tenour of perpetual decay, fall, renovation and progression' in 'the method of nature', developing by evolution rather than revolution. Plus ça change: just such a tree is the symbol of the UK Conservative and Unionist party, which continues, in the 21st century, to have problems with the politics of Europe. That Burke was one of the fathers of Romanticism is testimony to the essentially reactionary nature of the movement. John Ruskin shared this outlook, as this passage from *The Nature of Gothic* shows: 'It is evident that the chief feeling induced by woody country is one of reverence for its antiquity. There is a quiet melancholy about the decay of the patriarchal trunks, which is enhanced by the green and elastic vigor of the young saplings; the noble form of the forest aisles, and the subdued light which penetrates their entangled boughs, combine to add to the impression; and the whole character of the scene is calculated to excite conservative feeling. The man who could remain a radical in a wood country is a disgrace to his species' (Ruskin 1892). The contrast between the tranquil Arcadia of pastoral conservatism and the real world of staple crops that keep body and soul together couldn't be more stark.

## Chapter 15 The Tree Museum

The desk on which I write is piled up to here with books, most of which come in the category of New Nature Writing (Moran 2015). They were stacked there in the hope that they would give a picture of the current state of relations between people and plants. Disappointingly, for the most part they have turned out to be not very illuminating in themselves; but the debate about the genre within the community of literary critics does shed a kind of light, as well as being quite entertaining. Interestingly, some of the harshest appraisals have come from practitioners themselves. This is the poet Kathleen Jamie:

> 'What's that coming over the hill? A white, middle-class Englishman! A Lone Enraptured Male! From Cambridge! Here to boldly go, "discovering", then quelling our harsh and lovely and sometimes difficult land with his civilised lyrical words'.

New Nature Writing seems to provoke trenchant phrase-making in sceptics and sympathisers alike. As well as Lone Enraptured Male, we have Bogus Quest Narrative, Attractive Green Wash, Bourgeois Escapism - the list goes on (Smyth 2016). Something is troubling these people. What could it be?

This kind of literature clearly connects with the Romantic tradition, being preoccupied with individualism, emotional identification with the sublime in Nature and an elegiac yearning that sees the world as somehow having taken a wrong turning. You would be forgiven for supposing we are witnessing a kind of dawning Rousseauan realisation we've been on the wrong side in the conflict that's the subject of the present book. Evidence for growing regret is all around. It's there in the food culture of affluent urban consumerism. Veganism, 'clean eating', the paleo diet, the 'free-from' movement (eschewing gluten, dairy, nuts), all take issue with the fundamentals of the three million year old history of human-plant relationships. This is more than a mutiny: it feels like a defection to the other side. There is an appropriate German term in political philosophy that describes one in such a state of ambiguous radicalism: *Waldgänger*, walker in the woods (Horn 2004), someone estranged from the world at large a mensa et Thoreau, so to speak (with apologies to Myles na Gopaleen 1976).

Nothing better exemplifies the climate of apostasy than the rise within conservation biology of ecological restoration projects - often called 'rewilding' (Soulé and Noss 1998). The fugleman for rewilding in the UK is the environmentalist George Monbiot. His manifesto is *Feral* (Monbiot 2013). The subtitle, 'Searching for Enchantment on the Frontiers of Rewilding', places the book squarely in the Romantic tradition. What could be more sublime than the thrill of knowing that perilous impenetrable forests in the remotenesses might harbour beavers, wild boar, lynx, wolves, maybe even bears? Monbiot's work in the context of New Nature Writing has been critically examined in a review by Steven Poole (2013). He notes that Monbiot has a particular antipathy - hatred would not be too extreme a term - towards sheep. Like the author of the present work, Monbiot has lived in what he calls the 'sheep-scraped misery' of the Cambrian Desert. It is true that there are too many sheep and too few woods on the hills of Wales but, as Pryor (2003) and Poole (2013) point out, sheep have been here since the Neolithic Age, long before Saxons, Angles, Danes, Vikings, Romans, Normans and all the other incomers to these islands, and this entitles them to authentic cultural status. The pastoral tradition deserves respect, even if it can be hard on the environment (Blench 2001). Moreover, there's an ecological problem with getting humans and their beasts out of the picture so that the land can revert to primal forest. The favoured areas to be returned to wilderness tend to be those that were under glaciers until 10 Kya. The effect of glaciation was to scrape the terrain bare of everything except bedrock, and even now the land has only a thin skin of soil, an impoverished seed-bank and negligible seed-rain (Hill et al. 1992). Reversion to climax forest is going to take a long time without a helping hand (Vera 2000). Maybe the answer is a campaign of active planting. If so, what kind of wilderness is that? I saw a graffito recently in Canterbury, Kent, that says it all. A notice on a grassy bank near the city wall warned 'KEEP OFF. WILD FLOWERS PLANTED'. The annotation said 'IF THEY'RE PLANTED, THEY CAN'T BE WILD'. Cue Joni Mitchell (1970):

> They took all the trees
> And put them in a tree museum
> And they charged all the people
> A dollar and a half just to see 'em'

It's easy to forget that the writers of thoughtful books about nature and the environment (or even books like the present one), and the people who read them, are hardly representative of our species. Brute statistics tell us that their sensitivities are not those of humankind in general. While we fret about trees and grass and nature and culture and food and environment and climate and how to live a good sustainable life, and we buy books on deforestation from Amazon, the 7.5 billion people on the planet are overwhelmingly preoccupied with other things, and humanity is as busy slashing and burning as it's ever been. 'Treehugger' is not a term of admiration. When all is said and done, we remain in thrall to the grasses. Cereals, meat and dairy continue to be the foundations of the global human diet in the 21st century (Kearney 2010). Grass is everywhere, even in the heart of the concrete jungle (Haq 2011). Presidents and businessmen do deals while chasing golf balls around it (Watterson 2006). Millions watch sportspersons run, jump and kick on it (Dixon et al. 2015). Gardeners are slaves to grass (Robbins 2007; it's said that expenditure on machinery, chemicals and irrigation to maintain the lawns of Americans' yards vastly exceeds total inputs to Indian agriculture). And, as my friends Heather Ackroyd and Dan Harvey have shown, grass can make beautiful artworks (Britz 2009).

**Détente**

The artists Heather Ackroyd and Dan Harvey use natural materials, including grass and trees, to create works that engage with ecological and environmental concerns (Britz 2009, Ackroyd and Harvey 2011). *The Tree Ceremony* (2015) was exhibited in a symbolic staged event at the Musée National d'Histoire Naturelle and Jardin des Plantes, Paris. The artists comment: 'Framing the tree as 'actor' in an urban drama of increasing city temperatures, damaging flood waters, polluted air and bio-diversity loss, an evergreen oak holds centre-stage as impetus to green cities and towns world-wide'. In the war that the gives this book its title, perhaps a kind of redemption can be achieved through creative engagement with the living environment, as exemplified by Ackroyd and Harvey's work. More broadly, if humanity doesn't recalibrate its relationship with the natural world, the history of the relationship tells us that this world is fully capable of imposing a future according to its own implacable rules.

## AFTERWORD

Humans, for the most part, don't have a clue. Don't want or need one.
They're happy. They think they have a good bead on things.
 Agent K *Men in Black* (directed by Barry Sonnenfield 1997)

It is a clear night with an enormous full moon that appears to fill half the sky.
Must get a picture. Out comes the phone, point and shoot. And then the
disappointment - surely that tiny fuzzy blob in the image can't be the same moon
I was gazing at, with its craters and maria that seemed almost close enough to
touch. To see the moon as it really is, to experience one's surroundings without
assumptions and the baggage of preconception and psychology, it is
recommended that one bends over and looks at it through open legs (Coren
1992). Vladimir Nabokov (1983) said as much:

> 'If you have ever tried to stand and bend your head so as to look back
> between your knees, with your face turned upside down, you will see the
> world in a totally different light...Well, this trick of changing the vista, of
> changing the prism and the viewpoint, can be compared...to the kind of new
> twist through which you see a greener grass, a fresher world.'

*The War Between Trees and Grasses* describes the affinities and discordances
among humans and the plants they live with and on, as looked at upside-down
and backwards through the legs. Eating buttered toast may seem to establish
clear exploiter-exploited relationships between the human consumer, the grasses
that made the bread and fed the dairy animal, and the trees that burn so that we
can cook. But it only *seems* this way. The good scholar, whether of science,
humanities or the arts, should question everything, including this statement.

The lessons of evolution and ecology tell us that organism-organism interactions
are complex and reciprocal transactions, even when one of the organisms is *Homo
sapiens*, of the Godlike intellect and opposable thumb, and the others are, in every
sense, vegetables. Agency is the term used in the social sciences for this kind of
interactive give-and-take. Agency is not solely a human characteristic, and it is
much more multi-dimensional than simple exploitation. When it comes to

evolutionary fitness and survival, there is always agency (Sterelny 2001). For example, crop plants have been selected to minimise antinutritional and antibiotic defences. They cannot survive without human intervention (which includes both cultivation and adding back defence chemicals in the form of pesticides). Humans have exploited cereals; in turn, weeds such as darnel have exploited humans and cereals, endophytes exploit grasses; and everywhere in this mélange are genes behaving selfishly (Thomas et al. 2016). Humans, grasses, trees: a three million year-old web of agency.

This, then, has been my argument: our prehistoric ancestors unwittingly took the King's shilling (in the shape of polymers of glucose) and became combatants in grassland's age-old conflict with the forest. The war will likely never end, but this book ends here. Looking back to the start of this enterprise, I find myself reflecting sympathetically on the words of James Russell Lowell:

> In creating, the only hard thing's to begin;
> A grass-blade's no easier to make than an oak
> *A Fable For Critics* (1848)

# SOURCES

## Publications cited in the text

Abbo S, Gopher A, Rubin B, Lev-Yadun S. 2005. On the origin of Near Eastern founder crops and the 'dump-heap hypothesis'. Genetic Resources and Crop Evolution 52: 491–495.

Achebe C. 1958. *Things Fall Apart*. New York: Knopf.

Ackroyd H, Harvey D. 2011. Beuys' acorns. Antennae 17: 63-71.

Aerts R. 1995. The advantages of being evergreen. Trends in Ecology and Evolution. 10: 402-7.

Agnarsson I, Zhang JX. 2006. New species of *Anelosimus* (Araneae: Theridiidae) from Africa and Southeast Asia, with notes on sociality and color polymorphism. Zootaxa 1147: 1-34.

Albert RM, Bamford MK, Cabanes D. 2009. Palaeoecological significance of palms at Olduvai Gorge, Tanzania, based on phytolith remains. Quaternary International 193: 41–48.

Algeo TJ, Scheckler SE. 1998. Terrestrial-marine teleconnections in the Devonian: links between the evolution of land plants, weathering processes, and marine anoxic events. Philosophical Transactions of the Royal Society B 353: 113-130.

Ally D, Ritland K, Otto SP. 2010. Aging in a long-lived clonal tree. PLOS Biology 8(8): e1000454.

Ameisen JC. 2002. On the origin, evolution, and nature of programmed cell death: a timeline of four billion years. Cell Death and Differentiation 9: 367-93.

Amthor JS. 2010. From sunlight to phytomass: on the potential efficiency of converting solar radiation to phyto-energy. New Phytologist 188: 939–959.

Antón SC, Potts R, Aiello LC. 2014. Evolution of early *Homo*: An integrated biological perspective. Science 345: 1236828.

Arber A. 1950. *The Natural Philosophy of Plant Form*. Cambridge: University Press.

Archer JE, Thomas H, Marggraf Turley R. 2012. The Autumn King: remembering the land in King Lear. Shakespeare Quarterly 63: 518-543.

Archer J, Marggraf Turley R, Thomas H. 2014. *Food and the Literary Imagination*. London: Palgrave.

Armstrong GA. 1998. Greening in the dark: light-independent chlorophyll biosynthesis from anoxygenic photosynthetic bacteria to gymnosperms. Journal of Photochemistry and Photobiology B: Biology 43: 87-100.

Armstrong P. 2009. *Darwin's Luck: Chance and Fortune in the Life and Work of Charles Darwin*. London: Continuum.

Attenborough D. 1979. *Life on Earth*. London: Collins.

Axelsson E, Ratnakumar A, Arendt M-L, Maqbool K, Webster MT, Perloski M, Liberg O, Arnemo JM, Hedhammar Å, Lindblad-Toh K. 2013. The genomic signature of dog domestication reveals adaptation to a starch-rich diet. Nature 495: 360–364.

Bardgett RD, Mommer L, De Vries FT. 2014. Going underground: root traits as drivers of ecosystem processes. Trends in Ecology and Evolution 29: 692-699.

Barrett PM. 2000. Evolutionary consequences of dating the Yixian Formation. Trends in Ecology and Evolution 15: 99–103.

Bateman RM, DiMichele WA. 1994. Heterospory: the most iterative key innovation in the evolutionary history of the plant kingdom. Biological Reviews 69: 345–417.

Bateson W. 1894. *Materials for the Study of Variation (1992 edition)*. Baltimore: Johns Hopkins University Press.

Battistuzzi FU, Feijao A, Hedges SB. 2004. A genomic timescale of prokaryote evolution: insights into the origin of methanogenesis, phototrophy, and the colonization of land. BMC Evolutionary Biology 4: 44.

Baudena M, Dekker SC, Van Bodegom PM, Cuesta B, Higgins SI, Lehsten V, Reick CH, Rietkerk M, Scheiter S, Yin Z, Zavala MA. 2015. Forests, savannas, and grasslands: bridging the knowledge gap between ecology and Dynamic Global Vegetation Models. Biogeosciences 12: 1833-1848.

Beerling D. 2007. *The Emerald Planet: How plants changed Earth's history*. Oxford: University Press.

Bell AD. 1991. *Plant Form. An Illustrated Guide to Flowering Plant Morphology*. Oxford: University Press.

Benjamin B, Overton E. 1981. Prospects for mortality decline in England and Wales. Population Trends 23: 22-34.

Bennett J. 2010. *Sir Thomas Browne: 'A Man of Achievement in Literature'*. Cambridge University Press.

Bennici A. 2008. Origin and early evolution of land plants: Problems and considerations. Communicative and Integrative Biology 1: 212–218.

Beraldi-Campesi H. 2013. Early life on land and the first terrestrial ecosystems. Ecological Processes 2: 1.

Berger KG, Martin SM. 2000. Palm oil. In: Kiple KF, Ornelas KC (eds). *The Cambridge World History of Food Volume 1* pp. 397-410. Cambridge :University Press.

Berna F, Goldberg P, Horwitz LK, Brink J, Holt S, Bamford M, Chazan M. 2012. Microstratigraphic evidence of in situ fire in the Acheulean strata of Wonderwerk Cave, Northern Cape province, South Africa. Proceedings of the National Academy of Sciences, USA 109: E1215-1220.

Biraben JN. 2003. The rising numbers of humankind. Population and Societies. 394 (October): 1-4.

Blench R. 2001. *'You can't go home again'. Pastoralism in the new millennium*. London: ODI-FAO.

Bogin B. 1999. *Patterns of Human Growth 2nd edition*. p. 159. Cambridge: University Press.

Bond WJ. 2008. What limits trees in C₄ grasslands and savannas? Annual Review of Ecology, Evolution and Systematics 39: 641–659.

Bond WJ, Keeley JE. 2005. Fire as a global 'herbivore': the ecology and evolution of flammable ecosystems. Trends in Ecology and Evolution 20: 387-394.

Bond WJ, Woodward FI, Midgley GF. 2005. The global distribution of ecosystems in a world without fire. New Phytologist 165: 525-538.

Bouchenak-Khelladi YA, Verboom GA, Savolainen V, Hodkinson TR. 2010. Biogeography of the grasses (Poaceae): a phylogenetic approach to reveal evolutionary history in geographical space and geological time. Botanical Journal of the Linnaean Society 162: 543-557.

Boucher FC, Verboom GA, Musker S, Ellis AG. 2017. Plant size: a key determinant of diversification? New Phytologist 216: 24–31.

Box MS, Glover BJ. 2010. A plant developmentalist's guide to paedomorphosis: reintroducing a classic concept to a new generation. Trends in Plant Science 15: 241-246.

Boyce CK. 2010. The evolution of plant development in a paleontological context. Current Opinion in Plant Biology 13: 102-107.

Boyce CK, Fan Y, Zwieniecki MA. 2017. Did trees grow up to the light, up to the wind, or down to the water? How modern high productivity colors perception of early plant evolution. New Phytologist 215: 552–557.

Bradley R. 2005. *Ritual and Domestic Life in Prehistoric Europe*. Abingdon: Routledge.

Brand-Miller JC. 2003. Glycemic load and chronic disease. Nutrition Reviews 61: S49–55.

Britz S. 2009. Ackroyd and Harvey's grass. Antennae 10: 64-71.

Bröcker MJ, Watzlich D, Saggu M, Lendzian F, Moser J, Jahn D. 2010. Biosynthesis of (bacterio)chlorophylls: ATP-dependent transient subunit interaction and electron transfer of dark operative protochlorophyllide oxidoreductase. Journal of Biological Chemistry 285: 8268–8277.

Brunet M, Guy F, Pilbeam D, Mackaye HT, Likius A et al. 2002. A new hominid from the Upper Miocene of Chad, Central Africa. Nature 418: 145–151.

Buck-Sorlin G. 2013. Phytomer. In: *Encyclopedia of Systems Biology* pp. 1713-1714. New York: Springer.

Burger WC. 1998. The question of cotyledon homology in angiosperms. The Botanical Review 64: 356-371.

Campbell-Culver M. 2004. *The Origin of Plants*. London: Transworld.

Cantino PD, Doyle JA, Graham SW, Judd WS, Olmstead RG, Soltis DE, Soltis PS, Donoghue MJ. 2007. Towards a phylogenetic nomenclature of Tracheophyta. Taxon 56: 1E-44E.

Carlquist S. 2013. More woodiness/less woodiness: evolutionary avenues, ontogenetic mechanisms. International Journal of Plant Sciences 174: 964-991.

Carmody RN, Weintraub GS, Wrangham RW. 2011. Energetic consequences of thermal and nonthermal food processing. Proceedings of the National Academy of Sciences, USA 108: 19199–19203.

Cascales-Miñana B, Cleal CJ. 2014. The plant fossil record reflects just two great extinction events. Terra Nova 26: 195-200.

Catalan P, Chalhoub B, Chochois V, Garvin DF, Hasterok R, Manzaneda AJ, Mur LA, Pecchioni N, Rasmussen SK, Vogel JP, Voxeur A. 2014. Update on the genomics and basic biology of *Brachypodium*: International Brachypodium Initiative (IBI). Trends in Plant Science 19: 414-418.

Cerling TE, Harris JM. 1999. Carbon isotope fractionation between diet and bioapatite in ungulate mammals and implications for ecological and paleoecological studies. Oecologia 120: 347-363.

Cerling TE, Manthi FK, Mbua EN, Leakey LN, Leakey MG, Leakey RE, Brown FH, Grine FE, Hart JA, Kaleme P, Roche H. 2013. Stable isotope-based diet reconstructions of Turkana Basin hominins. Proceedings of the National Academy of Sciences, USA 110: 10501-10506.

Chakrabarty RK, Pervez S, Chow JC, Watson JG, Dewangan S, Robles J, Tian G. 2013. Funeral pyres in South Asia: Brown carbon aerosol emissions and climate impacts. Environmental Science and Technology Letters 1: 44-48.

Chanderbali AS, Berger BA, Howarth DG, Soltis PS, Soltis DE. 2016. Evolving ideas on the origin and evolution of flowers: new perspectives in the genomic era. Genetics 202: 1255-1265.

Christianson GE. 1972. Secret societies and agrarian violence in Ireland, 1790-1840. Agricultural History 46: 369-384.

Christin PA, Spriggs E, Osborne CP, Strömberg CA, Salamin N, Edwards EJ. 2014. Molecular dating, evolutionary rates, and the age of the grasses. Systematic Biology 63: 153-165.

Clarkson LA. 1960. The organization of the English leather industry in the late sixteenth and seventeenth centuries. The Economic History Review 13: 245-256.

Clauss M, Steuer P, Müller DWH, Codron D, Hummel J. 2013. Herbivory and body size: allometries of diet quality and gastrointestinal physiology, and implications for herbivore ecology and dinosaur gigantism. PLOS ONE 8: e68714.

Cobbett W. 1833. *Cobbett's Tour in Scotland and in the Four Northern Counties of England in the Autumn of the Year 1832*. London: William Cobbett.

Coen ES, Meyerowitz EM. 1991. The war of the whorls: genetic interactions controlling flower development. Nature 353: 31–37.

Collingham L. 2017. *The Hungry Empire: How Britain's Quest for Food Shaped the Modern World*. London: Bodley Head.

Conrad J. 1899. *Heart of Darkness*. London: Penguin (1973 edition).

Cooper A. 2014. Environment and the Natural World: annotated bibliography. In: King M (ed). *Oxford Bibliographies Online. Renaissance and Reformation*. Oxford: University Press (http://tinyurl.com/lwp5ou6 [accessed 4 October 2017]).

Coren S. 1992. The moon illusion: A different view through the legs. Perceptual and Motor Skills 75: 827-831.

Corner EJH. 1964. *The Life of Plants*. University of Chicago Press.

Crutzen PJ, Stoermer EF. 2000. The "Anthropocene". Global Change Newsletter 41: 17-18.

Cunniff J, Wilkinson S, Charles M, Jones G, Rees M, Osborne CP. 2014. Functional traits differ between cereal crop progenitors and other wild grasses gathered in the Neolithic fertile crescent. PLOS ONE 9: e87586.

Da Vinci L. 2016. *Notebooks Book VIII* 394. New Delhi: General Press.

Day ME, Greenwood MS, Diaz-Sala C. 2002. Age- and size-related trends in woody plant shoot development: regulatory pathways and evidence for genetic control. Tree Physiology 22: 507–513.

De Craene LR. 2016. Meristic changes in flowering plants: how flowers play with numbers. Flora - Morphology, Distribution, Functional Ecology of Plants 221: 22-37.

De Felice B. 2009. Transposable sequences in *Citrus* genome: role of mobile elements in the adaptation to stressful environments. Tree and Forestry Science and Biotechnology 3 (Special Issue 1): 79-86.

de Witte L, Stöcklin J. 2010. Longevity of clonal plants: why it matters and how to measure it. Annals of Botany 106: 859–870.

DeWoody J, Rowe CA, Hipkins VD, Mock KE. 2008. "Pando" lives: molecular genetic evidence of a giant aspen clone in Central Utah. Western North American Naturalist 68: 493-497.

Dietrich MR. 2003. Richard Goldschmidt: hopeful monsters and other 'heresies'. Nature Reviews in Genetics 4: 68–74.

Dixon S, Fleming P, James I, Carré M (eds). 2015. *The Science and Engineering of Sport Surfaces*. Abingdon: Routledge.

Dohn J, Dembélé F, Karembé M, Moustakas A, Amévor KA, Hanan NP. 2013. Tree effects on grass growth in savannas: competition, facilitation and the stress-gradient hypothesis. Journal of Ecology 101: 202–209.

Don GW. 1996. Goethe, Boretz, and the "Sensuous Idea". Perspectives of New Music 34: 124–139.

Doyle JA. 2012. Molecular and fossil evidence on the origin of angiosperms. Annual Review of Earth and Planetary Sciences 40: 301-26.

Doyle JA, Endress PK, Upchurch GR. 2008. Early Cretaceous monocots: a phylogenetic evaluation. Acta Musei Nationalis Pragae, Series B, Historia Naturalis 64: 59-87.

Dracopoli NC, Meisler MH. 1990. Mapping the human amylase gene cluster on the proximal short arm of chromosome 1 using a highly informative $(CA)_n$ repeat. Genomics 7: 97-102.

Drew BT, Ruhfel BR, Smith SA, Moore MJ, Briggs BG, Gitzendanner MA, Soltis PS, Soltis DE. 2014. Another look at the root of the angiosperms reveals a familiar tale. Systematic Biology 63: 368–382.

Drury S. 1992. Plants and pest control in England circa 1400–1700: a preliminary study. Folklore 103: 103-106.

Dujesiefken D, Fay N, de Groot J-W, de Berker N. 2016. *Trees – a Lifespan Approach*. Wroclaw: Fundacja EkoRozwoju.

Dulin MW, Kirchoff BK. 2010. Paedomorphosis, secondary woodiness, and insular woodiness in plants. Botanical Review 76: 405-490.

Dupont PY, Eaton CJ, Wargent JJ, Fechtner S, Solomon P, Schmid J, Day RC, Scott B, Cox MP. 2015. Fungal endophyte infection of ryegrass reprograms host metabolism and alters development. New Phytologist 208: 1227-1240.

Duru M, Therond O. 2015. Livestock system sustainability and resilience in intensive production zones: which form of ecological modernization? Regional Environmental Change 15: 1651-1665.

Edwards D. 1993. Cells and tissues in the vegetative sporophytes of early land plants. New Phytologist 125: 225-247.

Edwards D. 2003. Xylem in early tracheophytes. Plant, Cell and Environment 26: 57–72.

Edwards D, Li C-S, Raven JA. 2006. Tracheids in an early vascular plant: a tale of two branches. Botanical Journal of the Linnaean Society 150: 115-130.

Edwards EJ, Osborne CP, Strömberg CAE, Smith SA, C4 Grasses Consortium. 2010. The origins of $C_4$ grasslands: integrating evolutionary and ecosystem science. Science 328: 587-591.

Edwards EJ, Smith SA. 2010. Phylogenetic analyses reveal the shady history of $C_4$ grasses. Proceedings of the National Academy of Science, USA 107: 2532–2537.

Edwards J. 1789. *The Great Christian Doctrine of Original Sin Defended.* Boston, London: R Noble for J Murgatroyd.

Endress PK. 1987. Floral phyllotaxis and floral evolution. Botanische Jahrbücher für Systematik, Pflanzengeschichte und Pflanzengeographie 108: 417-438.

Endress PK, Doyle JA. 2009. Reconstructing the ancestral angiosperm flower and its initial specializations. American Journal of Botany 96: 22-66.

Eppich K. 2015. The decline and fall of the classic Maya city. In: Cherry JF, Rojas F (eds). *Archaeology for the People. Joukowsky Institute Publication No. 7* pp. 81-94. Oxford: Oxbow Books.

Esau K. 1960. *Anatomy of Seed Plants.* New York, London: Wiley.

Escamez S, Tuominen H. 2014. Programmes of cell death and autolysis in tracheary elements: when a suicidal cell arranges its own corpse removal. Journal of Experimental Botany 65: 1313-1321.

Evelyn, J. 1664. *Sylva, or A Discourse of Forest-Trees and the Propagation of Timber in His Majesty's Dominions*. London: John Martyn for the Royal Society.

Falkowski PG, Katz ME, Knoll AH, Quigg A, Raven JA, Schofield O, Taylor FJ. 2004. The evolution of modern eukaryotic phytoplankton. Science 305: 354-360.

Farquhar GD, Ehleringer JR, Hubick KT.1989. Carbon isotope discrimination and photosynthesis. Annual Review of Plant Biology 40: 503-537.

Farrelly D. 1984. *The Book of Bamboo*. San Francisco: Sierra Club Books.

Figueira TRS, Serrano GCM, Arruda P. 2008. Evolution of the genes encoding seed storage proteins in sugarcane and maize. Tropical Plant Biology 1: 108-119.

Fink S. 1983. The occurrence of adventitious and preventitious buds within the bark of some temperate and tropical trees. American Journal of Botany 70: 532-542.

Finlayson C. 2005. Biogeography and evolution of the genus *Homo*. Trends in Ecology and Evolution 20: 457-463.

Fisher CT, Fernández-Diaz JC, Cohen AS, Cruz ON, Gonzáles AM, Leisz SJ, Pezzutti F, Shrestha R, Carter W. 2016. Identifying ancient settlement patterns through LiDAR in the Mosquitia Region of Honduras. PLOS ONE 11: e0159890.

Fleischer R. 2005. *Out of the Inkwell: Max Fleischer and the Animation Revolution*. Lexington: University Press of Kentucky.

Flematti GR, Ghisalberti EL, Dixon KW, Trengove RD. 2004. A compound from smoke that promotes seed germination. Science 305: 977.

Folse HJ, Roughgarden J. 2012. Direct benefits of genetic mosaicism and intraorganismal selection: modeling coevolution between a long-lived tree and a short-lived herbivore. Evolution 66: 1091–1113.

Frank MH, Edwards MB, Schultz ER, McKain MR, Fei Z, Sørensen I, Rose JKC, Scanlon MJ. 2015. Dissecting the molecular signatures of apical cell-type shoot meristems from two ancient land plant lineages. New Phytologist 207: 893–904.

Frazer JG. 1993. *The Golden Bough*. London: Wordsworth.

Frenne P, Graae BJ, Rodríguez-Sánchez F, Kolb A, Chabrerie O, Decocq G, Kort H, Schrijver A, Diekmann M, Eriksson O, Gruwez R. 2013. Latitudinal gradients as natural laboratories to infer species' responses to temperature. Journal of Ecology 101: 784-795.

Friedman H. 2015. What on earth is the modern world-system? Foodgetting and territory in the modern era and beyond. Journal of World-Systems Research 26: 480-515.

Friedman J, Rubin MJ. 2015. All in good time: understanding annual and perennial strategies in plants. American Journal of Botany 102: 497-499.

Friedman WE. 2009. The meaning of Darwin's "abominable mystery". American Journal of Botany 96: 5-21.

Friedman WF. 2013. Mutants in our midst. Arnoldia 7: 1-14.

Friend WH. 1934. The origin of a superior red-fleshed grapefruit. Bud mutation of the Thompson variety of possible horticultural value. Journal of Heredity 25: 358-358.

Friis EM, Pedersen KR, Crane PR. 2006. Cretaceous angiosperm flowers: innovation and evolution in plant reproduction. Palaeogeography, Palaeoclimatology, Palaeoecology 232: 251-293.

Fukao T, Bailey-Serres J. 2004. Plant responses to hypoxia–is survival a balancing act? Trends in Plant Science 9: 449-456.

Fuller DQ. 2007. Contrasting patterns in crop domestication and domestication rates: recent archaeobotanical insights from the Old World. Annals of Botany 100: 903-924.

Ganiban R. 2013. The death and funeral rites in the Thebaid. In: Augoustakis A (ed). *Ritual and Religion in Flavian Epic* pp. 249-266. Oxford: University Press.

Gerrienne P, Bergamaschi S, Pereira E, Rodrigues MA, Steemans P. 2001. An Early Devonian flora, including *Cooksonia*, from the Paraná Basin (Brazil). Review of Palaeobotany and Palynology 116: 19-38.

Ghosh R, Naskar M, Bera S. 2011. Phytolith assemblages of grasses from the Sunderbans, India and their implications for the reconstruction of deltaic environments. Palaeogeography, Palaeoclimatology, Palaeoecology 311: 93-102.

Gibson W. 2010. *A Brief History of Britain 1660 – 1851*. London: Constable and Robinson.

Goldewijk KK, Beusen A. van Drecht G, de Vos M. 2011. The HYDE 3.1 spatially explicit database of human-induced global land-use change over the past 12,000 years. Global Ecology and Biogeography 20: 73–86.

González-Forero M, Faulwasser T, Lehmann L. 2017. A model for brain life history evolution. PLOS Computational Biology 13: e1005380.

Goodall M. 2016. What is a tree? Local Government Lawyer (http://bit.ly/2dYVP4o [accessed 4 October 2017]).

Gordenko NV, Broushkin AV. 2015. Ginkgoales: some problems of systematics and phylogeny. Paleontological Journal 49: 546-551.

Götmark F, Götmark E, Jensen AM. 2016. Why be a shrub? A basic model and hypotheses for the adaptive values of a common growth form. Frontiers in Plant Science 7: 1095.

Gould SJ, Eldredge N. 1977. Punctuated equilibria: the tempo and mode of evolution reconsidered. Paleobiology 3: 115–151.

Gowdy J, Krall L. 2013. The ultrasocial origin of the Anthropocene. Ecological Economics 95: 137-147.

Grann D. 2009. *The Lost City of Z: A Legendary British Explorer's Deadly Quest to Uncover the Secrets of the Amazon*. London: Simon and Schuster.

Graves R. 1948. *The White Goddess*. London: Faber and Faber (2010 edition).

Guan R, Zhao Y, Zhang H, Fan G, Liu X et al. 2016. Draft genome of the living fossil *Ginkgo biloba*. GigaScience 5: 49.

Gulbranson EL, Ryberg PE, Decombeix AL, Taylor EL, Taylor TN, Isbell JL. 2014. Leaf habit of Late Permian *Glossopteris* trees from high-palaeolatitude forests. Journal of the Geological Society 171: 493-507.

Hadley M. 2005. Nature to the fore: the early years of UNESCO's environmental programme, 1945-1965. In: Petitjean P, Zharov V, Glaser G, Richardson J, de Padirac B, Archibald G (eds.) *Sixty Years of Science at UNESCO 1945–2005* pp. 201-232. Paris: UNESCO.

Hagemann W. 1999. Towards an organismic concept of land plants: the marginal blastozone and the development of the vegetation body of selected frondose gametophytes of liverworts and ferns. Plant Systematics and Evolution 216: 81-133.

Hallé F. 1986. Modular growth in seed plants. Philosophical Transactions of the Royal Society B 313: 77-87.

Hallé F, Oldeman RAA, Tomlinson PB. 1978. *Tropical Trees and Forests: an Architectural Analysis*. Berlin: Springer-Verlag.

Hamant O, Moulia B. 2016. How do plants read their own shapes? New Phytologist 212: 333–337.

Hamilton J, Hedges REM, Robinson M. 2009. Rooting for pigfruit: pig feeding in Neolithic and Iron Age Britain compared. Antiquity 83: 998-1011.

Haq SM. 2011. Urban green spaces and an integrative approach to sustainable environment. Journal of Environmental Protection 2: 601-608.

Harbourne JB. 2014. *Introduction to Ecological Biochemistry* (4th edition, revised). London, San Diego: Academic Press.

Hardy K, Brand-Miller J, Brown KD, Thomas MG, Copeland L. 2015. The importance of dietary carbohydrate in human evolution. Quarterly Review of Biology 90: 251-268.

Harmand S, Lewis JE, Feibel CS, Lepre CJ, Prat S, Lenoble A, Boës X, Quinn RL, Brenet M, Arroyo A, Taylor N et al. 2015. 3.3-million-year-old stone tools from Lomekwi, West Turkana, Kenya. Nature 521: 310-315.

Harper JL. 1977. *The Population Biology of Plants*. London: Academic Press.

Harper JL, RosenBR, White J (eds). 1986. *The Growth and Form of Modular Organisms*. London: Royal Society.

Harris DR, Hillman GC (eds). 2014. *Foraging and Farming: The Evolution of Plant Exploitation. Volume 31 of Routledge Library Editions: Archaeology*. Oxford: Routledge.

Harrison JF, Kaiser A, VandenBrooks JM. 2010. Atmospheric oxygen level and the evolution of insect body size. Proceedings of the Royal Society of London B 277:1937-46.

Harrison RP. 1992. *Forests. The Shadow of Civilization*. University of Chicago Press.

Hartley SE, DeGabriel JL. 2016. The ecology of herbivore-induced silicon defences in grasses. Functional Ecology 30: 1311-22.

Hatcher P, Battey N. 2011. *Biological Diversity: Exploited and Exploiters*. Oxford: Wiley-Blackwell.

Hay WW, Migdisov A, Balukhovsky AN, Wold CN, Flögel S, Söding E. 2006. Evaporites and the salinity of the ocean during the Phanerozoic: Implications for climate, ocean circulation and life. Palaeogeography, Palaeoclimatology, Palaeoecology 240: 3-46.

Hayflick L, Moorhead P. 1961. The serial culture of human diploid cell strains. Experimental Cell Research 25: 585-621.

Heijden MG, Martin FM, Selosse MA, Sanders IR. 2015. Mycorrhizal ecology and evolution: the past, the present, and the future. New Phytologist 205: 1406-1423.

Helferich G. 2004. *Humboldt's Cosmos: Alexander Von Humboldt and the Latin American Journey that Changed the Way We See the World*. New York: Gotham Books.

Heller E. 1961. Goethe and the idea of scientific truth, in *The Disinherited Mind* pp. 8–9. Harmondsworth: Penguin.

Herendeen PS, Friis EM, Pedersen KR, Crane PR. 2017. Palaeobotanical redux: revisiting the age of the angiosperms. Nature Plants 3: 17015.

Hernandez-Aguilar RA. 2009. Chimpanzee nest distribution and site reuse in a dry habitat: Implications for early hominin ranging. Journal of Human Evolution 57: 350-364.

Hernandez-Aguilar RA, Moore J, Pickering TR. 2007. Savanna chimpanzees use tools to harvest the underground storage organs of plants. Proceedings of the National Academy of Sciences, USA 104: 19210-19213.

Hill MO, Evans DF, Bell SA. 1992. Long-term effects of excluding sheep from hill pastures in North Wales. Journal of Ecology 80: 1-13.

Hilton J, Bateman RM. 2006. Pteridosperms are the backbone of seed–plant phylogeny. Journal of the Torrey Botanical Society 133: 119–168.

Hochbach A, Schneider J, Röser M. 2015. A multi-locus analysis of phylogenetic relationships within grass subfamily Pooideae (Poaceae) inferred from sequences of nuclear single copy gene regions compared with plastid DNA. Molecular Phylogenetics and Evolution 87: 14-27.

Hofmann RR. 1989. Evolutionary steps of ecophysiological adaptation and diversification of ruminants: a comparative view of their digestive system. Oecologia 78: 443-57.

Holdgate GR, McLoughlin S, Drinnan AN, Finkelman RB, Willett JC Chiehowsky LA. 2005. Inorganic chemistry, petrography and palaeobotany of Permian coals in the Prince Charles Mountains, East Antarctica. International Journal of Coal Geology 63: 156-177.

Horn E. 2004. "Waldgänger", traitor, partisan: figures of political irregularity in West German postwar thought. CR: The New Centennial Review 4: 125-143.

Horton M. 2016. Meet Lidar: the amazing laser technology that's helping archaeologists discover lost cities. The Conversation (http://tinyurl.com/z6jppe8 [accessed 4 October 2017]).

Hsiao T. 2015. In defense of eating meat. Journal of Agricultural and Environmental Ethics 28: 277-291.

Hubbard RM, Bond BJ, Ryan MG. 1999. Evidence that hydraulic conductance limits photosynthesis in old *Pinus ponderosa* trees. Tree Physiology 19: 165-172.

Hughes NF. 1994. *The Enigma of Angiosperm Origins*. Cambridge: University Press.

Isely D. 2002. *One Hundred and One Botanists*. West Lafayette: Purdue University Press.

Ishii HT, Ford ED, Kennedy MC. 2007. Physiological and ecological implications of adaptive reiteration as a mechanism for crown maintenance and longevity. Tree Physiology 27: 455–462.

Janis C. 2007. An evolutionary history of browsing and grazing ungulates. In: Gordon IJ, Prins HHT (eds). *The Ecology of Browsing and Grazing* pp. 21-45. Berlin, Heidelberg: Springer.

Johnson SE, Abrams MD . 2009. Age class, longevity and growth rate relationships: protracted growth increases in old trees in the eastern United States. Tree Physiology 29: 1317–1328.

Jones R, Ougham H, Thomas H, Waaland S. 2013. *Molecular Life of Plants*. Chichester: Wiley.

Jones SS, Burke SV, Duvall MR. 2014. Phylogenomics, molecular evolution, and estimated ages of lineages from the deep phylogeny of Poaceae. Plant Systematics and Evolution. 300: 1421-1436.

Judson OP. 2017. The energy expansions of evolution. Nature Ecology and Evolution 1: 0138.

Jung CG. 1968. *Man and His Symbols*. New York: Dell Books.

Kaufman PB, Dayanandan P, Franklin CI, Takeoka Y. 1985. Structure and function of silica bodies in the epidermal system of grass shoots. Annals of Botany 55: 487-507.

Kays SJ, Nottingham SF. 2007. *Biology and Chemistry of Jerusalem Artichoke: Helianthus tuberosus L.* Baton Rouge: CRC Press.

Kearney J. 2010. Food consumption trends and drivers. Philosophical Transactions of the Royal Society B: Biological Sciences 365: 2793-2807.

Keenan MJ, Zhou J, Hegsted M, Pelkman C, Durham HA, Coulon DB, Martin RJ. 2015. Role of resistant starch in improving gut health, adiposity, and insulin resistance. Advances in Nutrition: An International Review Journal 6: 198-205.

Kenrick P, Strullu-Derrien C. 2014. The origin and early evolution of roots. Plant Physiology 166: 570-580.

Kenrick P, Wellman CH, Schneider H, Edgecombe GD. 2012. A timeline for terrestrialization: consequences for the carbon cycle in the Palaeozoic. Philosophical Transactions of the Royal Society B 367: 519-536.

Kikuzawa K, Lechowicz MJ. 2011. *Ecology of Leaf Longevity.* Tokyo: Springer Science and Business Media.

Kim SS, Douglas CJ. 2013. Sporopollenin monomer biosynthesis in Arabidopsis. Journal of Plant Biology 56: 1-6.

Kinmonth F. 2006. Ageing the yew - no core, no curve. International Dendrology Society Yearbook, 2005 pp. 41-46.

Kirkwood TBL. 2002. Evolution of ageing. Mechanisms of Ageing and Development 123: 737-745.

Klekowski EJ. 2003. Plant clonality, mutation, diplontic selection and mutational meltdown. Biological Journal of the Linnaean Society 79: 61–67.

Knauth LP, Kennedy MJ. 2009. Precambrian greening of the Earth. Nature 460: 728-732.

Kuldau GA, Liu J-S, White JF, Siegel MR, Schardl CL. 1997. Molecular systematics of Clavicipitaceae supporting monophyly of genus *Epichloë* and form genus *Ephelis*. Mycologia 89: 431–444.

Kuliukas A. 2014. Removing the "hermetic seal" from the aquatic ape hypothesis: waterside hypotheses of human evolution. Advances in Anthropology 4: 164-167.

Laden G, Wrangham R. 2005. The rise of the hominids as an adaptive shift in fallback foods: plant underground storage organs (USOs) and australopith origins. Journal of Human Evolution 49: 482-498.

Lanfear R, Ho SY, Davies TJ, Moles AT, Aarssen L, Swenson NG, Warman L, Zanne AE, Allen AP. 2013. Taller plants have lower rates of molecular evolution. Nature Communications 4: 1879.

Lang C. 2001. Deforestation in Vietnam, Laos and Cambodia. In: Vajpeyi DK (ed). *Deforestation, Environment, and Sustainable Development: A Comparative Analysis* pp. 111-137. Westport, London: Praeger

Langdon J. 1997. Umbrella hypotheses and parsimony in human evolution: a critique of the Aquatic Ape Hypothesis. Journal of Human Evolution 33: 479–494.

Lanner RM, Connor KF. 2001. Does bristlecone pine senesce? Experimental Gerontology 36: 675-685.

Larson G, Fuller DQ. 2014. The evolution of animal domestication. Annual Review of Ecology, Evolution, and Systematics 45: 115-136.

Lavender D. 1976. *California – a Bicentennial History*. New York, Nashville: Norton, AASLH.

Lee-Thorp JA, Sealy JC, Van Der Merwe NJ. 1989. Stable carbon isotope ratio differences between bone collagen and bone apatite, and their relationship to diet. Journal of Archaeological Science 16: 585-599.

Lennon B. 2009. Estimating the age of groups of trees in historic landscapes. Arboricultural Journal 32: 167-188.

Lenton TM, Dahl TW, Daines SJ, Mills BJW, Ozaki K, Saltzman MR, Porada P. 2016. Earliest land plants created modern levels of atmospheric oxygen. Proceedings of the National Academy of Sciences, USA 113: 9704–9709.

Leuchtmann A, Bacon CW, Schardl CL, White JF, Tadych M. 2014. Nomenclatural realignment of *Neotyphodium* species with genus *Epichloë*. Mycologia 106: 202–215.

Lieberman BS, Eldredge N. 2014. What is punctuated equilibrium? What is macroevolution? A response to Pennell et al. Trends in Ecology and Evolution 29: 185-186.

Lloyd PE, Dicken P. 1972. *Location in Space: a Theoretical Approach to Economic Geography*. New York: Harper and Row.

Loomis RS, Williams WA. 1963. Maximum crop productivity: an estimate. Crop Science 3: 67–72.

Looy CV. 2007. Extending the range of derived Late Paleozoic conifers: *Lebowskia* gen. nov. (Majonicaceae). International Journal of Plant Sciences 168: 957-972.

Looy CV. 2013. Natural history of a plant trait: branch-system abscission in Paleozoic conifers and its environmental, autecological, and ecosystem implications in a fire-prone world. Paleobiology 39: 235-252.

Lovelock JE; Margulis L. 1974. Atmospheric homeostasis by and for the biosphere: the Gaia hypothesis. Tellus. Series A. Stockholm: International Meteorological Institute. 26: 2–10.

Lucas WJ, Groover A, Lichtenberger R, Furuta K, Yadav SR, Helariutta Y, He XQ, Fukuda H, Kang J, Brady SM, Patrick JW. 2013. The plant vascular system: evolution, development and functions. Journal of Integrative Plant Biology 55: 294-388.

Lynch AJJ, Barnes RW, Cambecèdes J, Vaillancourt RE. 1998. Genetic evidence that *Lomatia tasmanica* (Proteaceae) is an ancient clone. Australian Journal of Botany 46: 25–33.

Mabberley DJ. 1974. Branching in pachycaul *Senecios*: The Durian Theory and the evolution of angiospermous trees and herbs. New Phytologist 73: 967-975.

MacDonald GM, Velichko AA, Kremenetski CV, Borisova OK, Goleva AA, Andreev AA, Cwynar LC, Riding RT, Forman SL, Edwards TW, Aravena R. 2000. Holocene treeline history and climate change across northern Eurasia. Quaternary Research 53: 302-311.

MacEvoy T. 2004. *Positive Impact Forestry*. Washington DC: Island Press.

Mackenzie G, Boa AN, Diego-Taboada A, Atkin SL, Sathyapalan T. 2015 Sporopollenin, the least known yet toughest natural biopolymer. Frontiers in Materials 2: article 66, pp. 1-5.

Maddison DR, Schulz K-S (eds). 2007. *The Tree of Life Web Project*. (http://tolweb.org [accessed 6 October 2017]).

Magallón S, Gómez-Acevedo S, Sánchez-Reyes LL, Hernández-Hernández T. 2015. A metacalibrated time-tree documents the early rise of flowering plant phylogenetic diversity. New Phytologist 207: 437–453.

Marggraf Turley R, Thomas H, Archer JE. 2010. A tragedy of idle weeds. Times Literary Supplement No. 5577 (19 February 2010) pp. 14-15.

Marlowe FM, Berbesque JC. 2009. Tubers as fallback foods and their impact on Hadza hunter–gatherers. American Journal of Physical Anthropology 140: 751–758.

Martens P. 1977. *Welwitschia mirabilis* and neoteny. American Journal of Botany 64: 916-920.

Martone PT, Estevez JM, Lu F, Ruel K, Denny MW, Somerville C, Ralph J. 2009. Discovery of lignin in seaweed reveals convergent evolution of cell-wall architecture. Current Biology 19: 169-175.

Maslow AH. 1943. A theory of human motivation. Psychological Review 50: 370–396.

Maurin O, Davies TJ, Burrows JE, Daru BH, Yessoufou K, Muasya AM, Bank M, Bond WJ. 2014. Savanna fire and the origins of the 'underground forests' of Africa. New Phytologist 204: 201-214.

McInnes L, Jones FA, Orme CD, Sobkowiak B, Barraclough TG, Chase MW, Govaerts R, Soltis DE, Soltis PS, Savolainen V. 2013. Do global diversity patterns of vertebrates reflect those of monocots?. PLOS ONE 8: e56979.

McKnight TD, Shippen DE. 2004. Plant telomere biology. The Plant Cell 16: 794-803.

McMahon TA, Kronauer RE. 1976. Tree structures: deducing the principle of mechanical design. Journal of Theoretical Biology 59: 443-466.

McGrew WC. 1992. *Chimpanzee Material Culture: Implications for Human Evolution.* Cambridge: University Press.

McGrew WC. 2013. Is primate tool use special? Chimpanzee and New Caledonian crow compared. Philosophical Transactions of the Royal Society of London B 368: 20120422.

McWethy DB, Higuera PE, Whitlock C, Veblen TT, Bowman DMJS, Cary GJ, Haberle SG, Keane RE, Maxwell BD, McGlone MS, Perry GLW, Wilmshurst JM, Holz A, Tepley AJ. 2013. A conceptual framework for predicting temperate ecosystem sensitivity to human impacts on fire regimes. Global Ecology and Biogeography 22: 900–912.

Mencuccini M, Martínez-Vilalta J, Vanderklein D, Hamid HA, Korakaki E, Lee S, Michiels B. 2005. Size-mediated ageing reduces vigour in trees. Ecology Letters 8: 1183–1190.

Merritt AD, Karn RC. 1977. The human $\alpha$-amylases. Advances in Human Genetics 8: 135-234.

Meyer-Berthaud B, Scheckler SE, Wendt J. 1999. *Archaeopteris* is the earliest known modern tree. Nature 398: 700-701.

Mihlbachler MC, Solounias N. 2006. Coevolution of tooth crown height and diet in oreodonts (Merycoidodontidae, Artiodactyla) examined with phylogenetically independent contrasts. Journal of Mammalian Evolution 13: 11–36.

Minamino R, Tateno M. 2014. Tree branching: Leonardo da Vinci's rule versus biomechanical models. PLOS ONE 9: e93535.

Mischi J. 2013. Contested rural activities: Class, politics and shooting in the French countryside. Ethnography 14: 64-84.

Mitchell J. 1970. Big Yellow Taxi. *Ladies of the Canyon*. Los Angeles: Reprise.

Moir A, Hindson T, Hills T, Haddlesey R. 2013. The exceptional yew trees of England, Scotland and Wales. Quarterly Journal of Forestry 107: 185-191.

Molloy P. 1985. *And They Blessed Rebecca: an Account of the Welsh Toll Gate Riots, 1839-44*. Llandysul: Gomer Press.

Monbiot G. 2013. *Feral: Searching for Enchantment on the Frontiers of Rewilding*. London: Allen Lane.

Moran J. 2014. A cultural history of the new nature writing. Literature and History 23: 49-63.

Munné-Bosch S. 2007. Aging in perennials. Critical Reviews in Plant Sciences 26: 123-138.

Muraki N, Nomata J, Ebata K, Mizoguchi T, Shiba T, Tamiaki H, Kurisu G, Fujita Y. 2010. X-ray crystal structure of the light-independent protochlorophyllide reductase. Nature 465: 110–114.

Nabokov V. 1983. *Lectures on Literature*. London: Picador.

na Gopaleen M. 1976. *The Various Lives of Keats and Chapman and The Brother*. London: Hart-Davis, MacGibbon.

Nic Eoin L. 2016. Geophytes, grasses and grindstones: replanting ideas of gathering in Southern Africa's Middle and Later Stone Ages. South African Archaeological Bulletin 71: 36–45.

Niklas KJ, Cobb ED, Kutschera U. 2016. Haeckel's Biogenetic Law and the land plant phylotypic stage. BioScience biw029.

Nisbet J. 1908. An Essay on the Life and Works of the Author. Introduction to the 4th edition of John Evelyn's *Sylva*, reprinted 1908. London: Doubleday.

Nogués-Bravo D, Rodríguez J, Hortal J, Batra P, Araújo MB. 2008. Climate change, humans, and the extinction of the woolly mammoth. PLOS Biology 6: e79.

O'Connell JF, Hawkes K, Jones NGJ. 1999. Grandmothering and the evolution of *Homo erectus*. Journal of Human Evolution 36: 461–485.

O'Hara K. 2010. *The Enlightenment: A Beginner's Guide*. London: Oneworld.

Onderdonk JJ, Ketcheson JW. 1972. A standardization of terminology for the morphological description of corn seedlings. Canadian Journal of Plant Science 52: 1003-1006.

Ortman SG, Cabaniss AH, Sturm JO, Bettencourt LM. 2015. Settlement scaling and increasing returns in an ancient society. Science Advances 1: e1400066.

Overton M. 1985. The diffusion of agricultural innovations in Early Modern England: Turnips and clover in Norfolk and Suffolk, 1580-1740. Transactions of the Institute of British Geographers 10: 205-221.

Palubicki W, Horel K, Longay S, Runions A, Lane B, Měch R. 2009. Self-organizing tree models for image synthesis. ACM Transactions on Graphics (TOG) 28: 58-67.

Parker ST, Gibson KR. 1977. Object manipulation, tool use and sensorimotor intelligence as feeding adaptations in Cebus monkeys and great apes. Journal of Human Evolution 6: 623-641.

Parsons JJ. 1962. The acorn–hog economy of the oak woodlands of southwestern Spain. Geographical Review 52: 211–235.

Pendergrast M. 2001. *Uncommon Grounds: The History of Coffee and How It Transformed Our World*. London: Texere.

Pennell MW, Harmon LJ, Uyeda JC. 2014. Is there room for punctuated equilibrium in macroevolution? Trends in Ecology and Evolution 29: 23-32.

Pennington RT, Hughes CE. 2014. The remarkable congruence of New and Old World savanna origins. New Phytologist 204: 4-6.

Perry GH, Dominy NJ, Claw KG, Lee AS, Fiegler H, Redon R, Werner J, Villanea FA, Mountain JL, Misra R, Carter NP. 2007. Diet and the evolution of human amylase gene copy number variation. Nature Genetics 39: 1256-1260.

Peters CR. 1987. Nut-like oil seeds: Food for monkeys, chimpanzees, humans, and probably ape-men. American Journal of Physical Anthropology 73: 333–363.

Pietsch KA, Ogle K, Cornelissen JH, Cornwell WK, Bönisch G, Craine JM, Jackson BG, Kattge J, Peltzer DA, Penuelas J, Reich PB. 2014. Global relationship of wood and leaf litter decomposability: the role of functional traits within and across plant organs. Global Ecology and Biogeography 23: 1046-1057.

Pigg KB, Trivett ML. 1994. Evolution of the glossopterid gymnosperms from Permian Gondwana. Journal of Plant Research 107: 461-477.

Piperno DR, Sues HD. 2005. Dinosaurs dined on grass. Science 310: 1126-1128.

Plackett ARG, Coates JC. 2016. Life's a beach – the colonization of the terrestrial environment. New Phytologist 212: 831–835.

Plumstead EP. 1958. The habit of growth of Glossopteridae. Transactions and Proceedings of the Geological Society of South Africa 61: 81–96.

Poinar G, Alderman ST, Wunderlich JO. 2015. One hundred million year old ergot: psychotropic compounds in the Cretaceous? Palaeodiversity 8: 13-19.

Pollock CJ. 1986. Fructans and the metabolism of sucrose in vascular plants. New Phytologist 104: 1-24.

Poole S. 2013. Is our love of nature writing bourgeois escapism? The Guardian (https://www.theguardian.com/books/2013/jul/06/nature-writing-revival [accessed 4 October 2017]).

Popper ZA, Michel G, Hervé C, Domozych DS, Willats WG, Tuohy MG, Kloareg B, Stengel DB. 2011. Evolution and diversity of plant cell walls: from algae to flowering plants. Annual Review of Plant Biology 62: 567-590.

Porter R. 2001. *The Enlightenment*. Basingstoke: Palgrave Macmillan.

Powell MJ. 2005. Popular disturbances in late eighteenth-century Ireland: the origins of the Peep of Day Boys. Irish Historical Studies 34: 249-265.

Prasad V, Strömberg CAE, Leaché AD, Samant B, Patnaik R, Tang L, Mohabey DM, Ge S, Sahni A. 2011. Late Cretaceous origin of the rice tribe provides evidence for early diversification in Poaceae. Nature Communications 2: 480.

Primary Teachers' Notes 2014-15. *Mr and Mrs Andrews, c.1750. oil on canvas 69.8 x 119.4 cm. Thomas Gainsborough*. London: The National Gallery.

Prothero D. 2013. *Rhinoceros Giants: The Palaeobiology of Indricotheres*. Bloomington: Indiana University Press.

Prusinkiewicz P, Runions A. 2012. Computational models of plant development and form. New Phytologist 193: 549–569.

Pryor F. 2003. *Britain BC*. London: HarperCollins.

Raunkiær C. 1937. *Plant Life Forms*. London: Clarendon Press

Raup DM. 1994. The role of extinction in evolution. Proceedings of the National Academy of Science, USA 91: 6758–6763.

Raven JA. 1986. Evolution of plant life forms. In: Givnish TJ (ed.) *On the Economy of Plant Form and Function* pp. 421–492. New York: Cambridge University Press.

Raven JA. 2005. Cellular location of starch synthesis and evolutionary origin of starch genes. Journal of Phycology 41: 1070-1072.

Raven JA, Edwards D. 2001. Roots: evolutionary origins and biogeochemical significance. Journal of Experimental Botany 52: 381–401.

Rees PM. 2002. Land-plant diversity and the end-Permian mass extinction. Geology 30: 827-830.

Rees PM, Gibbs MT, Ziegler AM, Kutzbach JE, Behling PJ. 1999. Permian climates: evaluating model predictions using paleobotanical data. Geology 27: 891-894.

Richards E. 2000. *The Highland Clearances*. Edinburgh: Birlinn Press.

Robbins P. 2007. *Lawn People: How Grasses, Weeds, and Chemicals Make Us Who We are*. Philadelphia: Temple University Press.

Rothwell GW, Wyatt SE, Tomescu AM. 2014. Plant evolution at the interface of paleontology and developmental biology: an organism-centered paradigm. American Journal of Botany 101: 899-913.

Rowe JS. 1983. Concepts of fire effects on plant individuals and species. In: Wein RW, MacLean DA (eds). *The Role of Fire in Northern Circumpolar Ecosystems* pp. 135-154. New York: Wiley.

Royer DL, Berner RA, Montañez IP, Tabor NJ, Beerling DJ. 2004. $CO_2$ as a primary driver of phanerozoic climate. GSA today 14: 4-10.

Rozema J, Björn LO, Bornman JF, Gaberščik A, Häder DP, Trošt T, Germ M, Klisch M, Gröniger A, Sinha RP, Lebert M. 2002. The role of UV-B radiation in aquatic and terrestrial ecosystems—an experimental and functional analysis of the evolution of UV-absorbing compounds. Journal of Photochemistry and Photobiology B: Biology 66: 2-12.

Rudall PJ, Buzgo M. 2004. Evolutionary history of the monocot leaf. In: Cronk QCB, Bateman RM, Hawkins JA (eds). *Developmental Genetics and Plant Evolution* pp. 431-458. London, New York: Taylor and Francis.

Ruddiman WF. 2003. The anthropogenic greenhouse era began thousands of years ago. Climatic Change 61: 261–293.

Ruddiman WF, Ellis EC, Kaplan JO, Fuller DQ. 2015. Defining the epoch we live in. Science 348: 38–39.

Ruddiman WF, Thomson JS. 2001. The case for human causes of increased atmospheric CH4 over the last 5000 years. Quaternary Science Reviews 20: 1769-1777.

Ruff AR. 2015. *Arcadian Visions: Pastoral Influences on Poetry, Painting and the Design of Landscape*. Oxford: Windgather Press.

Ruskin J. 1892. The nature of Gothic: a chapter of *The Stones of Venice, Volume II*. London and Orpington: George Allen.

Rydin C, Friis EM. 2010. A new Early Cretaceous relative of Gnetales: *Siphonospermum simplex* gen. et sp. nov. from the Yixian Formation of Northeast China. BMC Evolutionary Biology 10: 183.

Saikkonen K, Gundel PE, Helander M. 2013. Chemical ecology mediated by fungal endophytes in grasses. Journal of Chemical Ecology. 39: 962-968.

Saguaro S, Thacker DC. 2013. Tolkien and trees. In: Hunt P (ed). *J.R.R. Tolkien (New Casebooks)* pp. 139-155. Basingstoke: Palgrave Macmillan.

Salguero-Gomez R; Shefferson RP; Hutchings MJ. 2013. Plants do not count... or do they? New perspectives on the universality of senescence. Journal of Ecology 101: 545-554.

Sanders H, Rothwell GW, Wyatt S. 2007. Paleontological context for the developmental mechanisms of evolution. International Journal of Plant Sciences 168: 719-728.

Sauquet H, von Balthazar M, Magallón S, Doyle JA, Endress PK et al. 2017. The ancestral flower of angiosperms and its early diversification. Nature Communications 8: 16047.

Scarpella E, Meijer AH. 2004. Pattern formation in the vascular system of monocot and dicot plant species. New Phytologist 164: 209-242.

Schama S. 1995. *Landscape and Memory*. New York: Knopf.

Schardl CL, Young CA, Faulkner JR, Florea S, Pan J. 2012. Chemotypic diversity of *Epichloae*, fungal symbionts of grasses. Fungal Ecology 5: 331–344.

Scheiter S, Higgins SI, Osborne CP, Bradshaw C, Lunt D, Ripley BS, Taylor LL, Beerling DJ. 2012. Fire and fire-adapted vegetation promoted $C_4$ expansion in the late Miocene. New Phytologist 195: 653–666.

Scott W. 1818. *Heart of Midlothian*. Edinburgh: Constable.

Shipley GP, Kindscher K. 2016. Evidence for the Paleoethnobotany of the Neanderthal: a review of the literature. Scientifica 8927654.

Schweingruber FH. 1988. *Tree Rings: Basics and Applications of Dendrochronology.* Dordrecht: Kluwer.

Semaw S. 2000. The world's oldest stone artefacts from Gona, Ethiopia: their implications for understanding stone technology and patterns of human evolution between 2.6–1.5 million years ago. Journal of Archaeological Science 27: 1197–1214.

Seuss E. 1885. *Das Antlitz der Erde* (*The Face of the Earth*), vol. 1, p 768. Leipzig: Freytag.

Shapiro JA. 2013. Rethinking the (im) possible in evolution. Progress in Biophysics and Molecular Biology 111: 92-96.

Shelton D, Vaz A, Castillo BH, Nassar CC, Bello LJ, Colleoni P, Proaño J et al. 2013. *Indigenous Peoples in Voluntary Isolation and Initial Contact.* Pamplona, Copenhagen: IPES-IWGIA (http://bit.ly/2pKgL4d [accessed 4 October 2017]).

Sibley DA. 2009. *The Sibley Guide to Trees*. New York: Knopf.

Siegel MR, Latch GC, Johnson MC. 1987. Fungal endophytes of grasses. Annual Review of Phytopathology 25: 293-315.

Silvestro D, Cascales-Miñana B, Bacon CD, Antonelli A. 2015. Revisiting the origin and diversification of vascular plants through a comprehensive Bayesian analysis of the fossil record. New Phytologist 207: 425-36.

Sköld HN, Asplund MD, Wood CA, Bishop JDD. 2011. Telomerase deficiency in a colonial ascidian after prolonged asexual propagation. Journal of Experimental Zoology (B. Molecular and Developmental Evolution) 316: 276–283.

Skottsberg C (ed). 1953. *The Natural History of Juan Fernández and Easter Island. Volume 2 Botany*. Uppsala: Almqvist and Wiksells Boktryckeri.

Smyth R. 2016. The cult of nature writing. New Humanist, Spring edition pp. 50-53.

Sokoloff DD, Rudall PJ, Bateman RM, Remizowa MV. 2015. Functional aspects of the origin and subsequent evolution of cotyledons in seed plants. Botanica Pacifica 4: 1-13.

Soltis DE, Bell CD, Kim S, Soltis PS. 2008. Origin and early evolution of angiosperms. Annals of the New York Academy of Sciences 1133: 3-25.

Soltis P, Soltis D, Edwards C. 2005. Angiosperms. Flowering Plants. Version 03. http://tolweb.org/Angiosperms/20646/2005.06.03 in The Tree of Life Web Project, http://tolweb.org/ [accessed 6 October 2017].

Soulé M, Noss R. 1998. Rewilding and biodiversity: complementary goals for continental conservation. Wild Earth 8: 19–28.

Spicer R, Groover A. 2010. Evolution of development of vascular cambia and secondary growth. New Phytologist 186: 577–592.

Sponheimer M, Alemseged Z, Cerling TE, Grine FE, Kimbel WH, Leakey MG, Lee-Thorp JA, Manthi FK, Reed KE, Wood BA, Wynn JG. 2013. Isotopic evidence of early hominin diets. Proceedings of the National Academy of Sciences, USA. 110: 10513-10518.

Stanford CB. 1998. The social behavior of chimpanzees and bonobos: empirical evidence and shifting assumptions. Current Anthropology 39: 399–420.

Stebbins GL. 1981. Coevolution of grasses and herbivores. Annals of the Missouri Botanical Garden 68: 75-86.

Steffes M. 2016. Medieval wildernesses and King Lear: heath, forest, desert. Exemplaria 28: 230-247.

Stein WE, Mannolini F, Hernick LV, Landing E, Berry CM. 2007. Giant *cladoxylopsid* trees resolve the enigma of the Earth's earliest forest stumps at Gilboa. Nature 446: 904-907.

Steinfeld H, Gerber P, Wassenaar T, Castel V, Rosales M, de Haan C. 2006. *Livestock's Long Shadow: Environmental Issues and Options*. Rome: FAO.

Sterelny K. 2001. *The Evolution of Agency and Other Essays*. Cambridge: University Press.

Sterelny K. 2007. *Dawkins vs Gould: Survival of the Fittest*. Cambridge: Icon Books.

Steur H. 2016. *Hans' Paleobotany Pages*. (https://steurh.home.xs4all.nl/home.html#inhoud [accessed 4 October 2017]).

Strömberg CA. 2011. Evolution of grasses and grassland ecosystems. Annual Review of Earth and Planetary Sciences 39: 517-544.

Takahata N, Satta Y. 1997. Evolution of the primate lineage leading to modern humans: phylogenetic and demographic inferences from DNA sequences. Proceedings of the National Academy of Science, USA 94: 4811–4815.

Tambling J. 2005. *Blake's Night Thoughts*. Basingstoke: Palgrave Macmillan.

Taylor TN, Taylor EL, Krings M. 2009. *Paleobotany: The Biology and Evolution of Fossil Plants (2nd edition)*. Cambridge MA: Academic Press.

Teichmann T, Muhr M. 2015. Shaping plant architecture. Frontiers in Plant Science 6: 233.

Theissen G. 2009. Saltational evolution: hopeful monsters are here to stay. Theory in Biosciences 128: 43-51.

Theissen G, Melzer R. 2007 Molecular mechanisms underlying origin and diversification of the angiosperm flower. Annals of Botany 100: 603–619.

Thomas BA, Cleal C J. 1999. Abscission in the fossil record. In: Kurmann MR, Hemsley AR (eds). *Evolution of Plant Architecture* pp. 183-203. Kew: Royal Botanic Gardens.

Thomas H. 1994. Aging in the plant and animal kingdoms – the role of cell death. Reviews in Clinical Gerontology 4: 5-20.

Thomas H. 1998. Air today – gone tomorrow. New Phytologist 139: 225-229.

Thomas H. 2013. Senescence, ageing and death of the whole plant. New Phytologist 197: 696–711.

Thomas H. 2016. *Senescence*. Aberystwyth: Thomas (www.plantsenescence.org).

Thomas H. 2017. A green epoch in the evolutionary history of biological energy sources. Nature Ecology and Evolution 1: doi:10.1038/s41559-017-0302-8.

Thomas H, Archer JE, Marggraf Turley R. 2011. Evolution, physiology and phytochemistry of the psychotoxic arable mimic weed darnel (*Lolium temulentum* L). Progress in Botany 72: 73–104.

Thomas H, Archer JE, Marggraf Turley R. 2016. Remembering darnel, a forgotten plant of literary, religious and evolutionary significance. Journal of Ethnobiology 36: 29-44

Thomas H, Huang L, Young M, Ougham H. 2009. Evolution of plant senescence. BMC Evolutionary Biology 9: 163.

Thomas H, Sadras VO. 2001. The capture and gratuitous disposal of resources by plants. Functional Ecology 15: 3-12.

Tiffney BH, Niklas KJ. 1985. Clonal growth in land plants, a paleobotanical perspective. In: Jackson JBC, Buss LW, Cook RE (eds). *Population Biology and Evolution of Clonal Organisms*. pp. 35-66. New Haven: Yale University Press.

Tomlinson PB, Zimmermann MH. 1969. Vascular anatomy of monocotyledons with secondary growth - an introduction. Journal of the Arnold Arboretum 50: 159-79.

Toynbee A. 1976. *Mankind and Mother Earth*. Oxford: University Press.

Traverse A. 1988. Plant evolution dances to a different beat: plant and animal evolutionary mechanisms compared. Historical Biology 1: 277-301.

Tudge C. 2004. *So Shall We Reap: What's Gone Wrong with the World's Food - and How to Fix it*. London: Penguin.

Tully J. 2011. *The Devil's Milk: A Social History of Rubber*. New York: University Press.

Turnbull JW. 2009. Tree domestication and the history of plantations. In: Squires VR (ed). *The Role of Food, Agriculture, Forestry and Fisheries in Human Nutrition Vol II*. pp. 48–74. Encyclopedia of Life Support Systems, Paris: UNESCO-EOLSS.

Ungar PS, Sponheimer M. 2011. The diets of early hominins. Science 334: 190-193.

Van Noordwijk M, Lawson G, Hairiah K, Wilson J. 2015. Root distribution of trees and crops: competition and/or complementarity. In: Ong CK, Black C, Wilson J (eds). *Tree–Crop Interactions (2nd edition): Agroforestry in a Changing Climate* pp. 221-257. Wallingford: CABI

van Thünen JH. 1966. *The Isolated State*. Translated from the 1826 original by Wartenberg CM. Oxford: Pergamon.

Vavilov NI, Dorofeev VF. 1992. *Origin and Geography of Cultivated Plants*. Cambridge University Press.

Vera FWM. 2000. *Grazing Ecology and Forest History*. Wallingford: CABI.

Verdú M. 2002. Age at maturity and diversification in woody angiosperms. Evolution 56: 1352–1361.

Vicari M, Bazely DR. 1993. Do grasses fight back? The case for antiherbivore defences. Trends in Ecology and Evolution 8: 137-141.

Vos J, Evers JB, Buck-Sorlin GH, Andrieu B, Chelle M, de Visser PHB. 2010. Functional–structural plant modelling: a new versatile tool in crop science. Journal of Experimental Botany 61: 2101-2115.

Walbot V, Evans MMS. 2003. Unique features of the plant life cycle and their consequences. Nature Reviews Genetics 4: 369-379.

Waller R. 1962. *Prophet of the New Age. The Life and Thought of Sir George Stapledon FRS*. London: Faber.

Wang S, Li C, Copeland L, Niu Q, Wang S. 2015. Starch retrogradation: A comprehensive review. Comprehensive Reviews in Food Science and Food Safety 14: 568-585.

Warde P. 2006. Fear of wood shortage and the reality of the woodland in Europe, c. 1450–1850. History Workshop Journal 62: 28-57.

Warren JM. 2015. *The Nature of Crops. How we came to eat the plants we do.* Wallingford: CABI.

Waters CN, Zalasiewicz J, Summerhayes C, Barnosky AD, Poirier C. 2016. The Anthropocene is functionally and stratigraphically distinct from the Holocene. Science. 351: aad2622.

Watterson JS. 2006. *The Games Presidents Play: Sports and the Presidency*. Baltimore: JHU Press.

Waugh E. 1934. *A Handful of Dust*. London: Chapman and Hall.

Weber E. 2003. *Invasive Plant Species of the World: a Reference Guide to Environmental Weeds*. Wallingford: CABI.

Weismann AFL. 1893. *The Germ-plasm: A Theory of Heredity*. New York: Scribner.

Wellman CH. 2010. The invasion of the land by plants: when and where? New Phytologist 188: 306-309.

Weng JK, Chapple C. 2010. The origin and evolution of lignin biosynthesis. New Phytologist 187: 273-285.

Wiens JJ. 2016. Climate-related local extinctions are already widespread among plant and animal species. PLOS Biology 14: e2001104.

Wiersum KF. 1997. From natural forest to tree crops, co-domestication of forests and tree species, an overview. Netherlands Journal of Agricultural Science 45: 425-438.

Wikström N, Savolainen V, Chase MW. 2001. Evolution of the angiosperms: calibrating the family tree. Proceedings of the Royal Society B. 268: 2211-2220.

Williams JH. 2009. *Amborella trichopoda* (Amborellaceae) and the evolutionary developmental origins of the angiosperm progamic phase. American Journal of Botany 96: 144-65.

Willis KJ, McElwain JC. 2014. *The Evolution of Plants (2nd edition)*. Oxford: University Press.

Wilson SD. 1998. Competition between grasses and woody plants. In: Cheplick GP (ed). *Population Biology of Grasses* pp 231-254. Cambridge: University Press.

Wohl CG. 2007. Scientist as detective: Luis Alvarez and the pyramid burial chambers, the JFK assassination, and the end of the dinosaurs. American Journal of Physics 75: 968.

Wrangham RW. 2009. *Catching Fire: How Cooking Made Us Human*. New York: Basic Books.

Wyka TP, Oleksyn J. 2014. Photosynthetic ecophysiology of evergreen leaves in the woody angiosperms-a review. Dendrobiology 72: 3-27.

Wynn T, Hernandez-Aguilar RA, Marchant LF, Mcgrew WC. 2011. "An ape's view of the Oldowan" revisited. Evolutionary Anthropology 20: 181-197.

Xu B, Ohtani M, Yamaguchi M, Toyooka K, Wakazaki M et al. 2014. Contribution of NAC transcription factors to plant adaptation to land. Science 343: 1505-1508.

Xue J, Deng Z, Huang P, Huang K, Benton MJ, Cui Y, Wang D, Liu J, Shen B, Basinger JF, Hao S. 2016. Belowground rhizomes in paleosols: the hidden half of an Early Devonian vascular plant. Proceedings of the National Academy of Sciences, USA 8: 201605051.

Young AM. 1994. *The Chocolate Tree: a Natural History of Cacao*. Washington DC: Smithsonian Institution Press.

Zanis MJ, Soltis DE, Soltis PS, Mathews S, Donoghue MJ. 2002. The root of the angiosperms revisited. Proceedings of the National Academy of Sciences, USA 99: 6848-6853.

Zavaleta-Mancera HA, Franklin KA, Ougham HJ, Thomas H, Scott IM. 1999. Regreening of senescent *Nicotiana* leaves. I. Reappearance of NADPH-protochlorophyllide oxidoreductase and light-harvesting chlorophyll a/b-binding protein. Journal of Experimental Botany 50: 1677-1682.

Zeeman SC, Kossmann J, Smith AM. 2010. Starch: its metabolism, evolution, and biotechnological modification in plants. Annual Review of Plant Biology 61: 209-234.

Zhao P, Begcy K, Dresselhaus T, Sun MX. 2017. Does early embryogenesis in eudicots and monocots involve the same mechanism and molecular players?. Plant Physiology 173: 130-142.

Zhong B, Liu L, Yan Z, Penny D. 2013. Origin of land plants using the multispecies coalescent model. Trends in Plant Science 18:492-495.

Zhou ZY. 2009. An overview of fossil Ginkgoales. Palaeoworld 18: 1–22.

Zuk M. 2013. *Paleofantasy: What Evolution Really Tells Us about Sex, Diet, and How We Live.* New York: Norton.

Zwieniecki MA, Boyce CK. 2014. Evolution of a unique anatomical precision in angiosperm leaf venation lifts constraints on vascular plant ecology. Proceedings of the Royal Society B 281: 20132829.

## Sources of illustrations

Cover design by author and Debbie Maizels. Blazing trees copyright Loopall (reproduced under standard licence from www.123rf.com/ profile_Loopall).

Page 3. Amazonian forest (http://foundtheworld.com/amazon-rainforest/ [accessed 2 October 2017]; public domain).

Page 8. Ordovician scene, by Jose Bonner (https://commons.wikimedia.org/ wiki/File:Ordovician_Land_Scene.jpg. [accessed 2 October 2017]; public domain, under Creative Commons licence). Devonian forest biome, by Hussan Sheikh (www.slideshare.net/hussanara/ancient-forests-stabilized-earths-co2-and-climate [accessed 2 October 2017]; public domain).

Page 12. Micrographs of vascular tissues and tracheid cells, reproduced with kind permission of Professor Dianne Edwards and New Phytologist.

Page 13. *Archaeopteris* and *Glossopteris* copyright Corey A Ford (rwww.123rf.com/profile_catmando). Fossilised leaf bed of *Archaeopteris* from http://tinyurl.com/y7x5v4t3 [accessed 1 October 2017]; public domain, under Creative Commons licence. Photograph of fossilised *Glossopteris* leaf bed by Ghedoghedo (http://tinyurl.com/ya7ln6rh [accessed 1 October 2017]; public domain, under Creative Commons licence).

Page 18. Modular structure of idealised dicot and monocot plant bodies from the open access article by Teichmann and Muhr (2015), distributed under the terms of the Creative Commons Attribution License (CC BY).

Page 22. Llangernyw Yew, by Emgaol (https://commons.wikimedia.org/wiki/ File:The_Llangernyw_yew.jpg [accessed 2 October 2017]; public domain, under Creative Commons licence). King's Holly (*Lomatia tasmanica*) from http://www.abc.net.au/news/2014-08-09/rare-and-endangered-tasmanian-plant-lomatia/5660022 [accessed 2 October 2017], public domain.

Page 23. Shoot apical meristem, reproduced from *The Molecular Life of Plants*, with kind permission of the publisher, Wiley. Early Devonian shoot apex from http://tinyurl.com/mmzoffb [accessed 2 October 2017], kindly provided by Dr Philippe Gerrienne. Ruby grapefruit from http://green-food.net/en/ecological-star-ruby-grapefruit-season-begins-at-green-food/ [accessed 2 October 2017], public domain.

Page 27. Fossil ginkgophyte leaf from Tangopaso (https://commons.wikimedia.org/wiki/File:Ginkgo_biloba_MacAbee_BC.jpg [accessed 2 October 2017]; public domain under Creative Commons licence). Green *Ginkgo* leaves by James Field (https://commons.wikimedia.org/wiki/File:Ginkgo_Biloba_Leaves_-_Black_Background.jpg [accessed 2 October 2017]; public domain under Creative Commons licence).

Page 28. *The Large Piece of Turf*, by Albrecht Dürer (1503). From https://commons.wikimedia.org/wiki/File:Albrecht_D%C3%BCrer_-_The_Large_Piece_of_Turf,_1503_-_Google_Art_Project.jpg [accessed 2 October 2017], public domain.

Page 33. Public domain version of Betty Boop image (U.S. Patent D86,224), (https://commons.wikimedia.org/w/index.php?curid=38197116 [accessed 2 October 2017]). Photograph of giant tree *Senecio* by Stig Nygaard, reproduced from Wikimedia under Creative Commons (https://commons.wikimedia.org/wiki/File:Giant_Tree_Groundsel.jpg [accessed 2 October 2017]). Photograph of *Robinsonia* from http://www.wondermondo.com/Countries/SA/Chile/Valparaiso/RobinsonCrusoeForest.htm [accessed 2 October 2017], public domain.

Page 38. *Siphonospermum* fossil, from Rydin C, Friis EM. 2010. A new Early Cretaceous relative of Gnetales: *Siphonospermum simplex* gen. et sp. nov. from the Yixian Formation of Northeast China. BMC Evolutionary Biology 10: 183 (open access publication under Creative Commons licence). Image of fossilised grass seed and flower generously provided by Hans Steur.

Page 45. Tree image copyright Andrey Kryuchkov (www.123rf.com/profile_varunalight); grass image copyright Denys Prokofyev (www.123rf.com/profile_denisnata); tree rings copyright Peter Hermes Furian (www.123rf.com/profile_peterhermesfurian); cross -section of grass stem public domain (http://tinyurl.com/yb7r2j3a [accessed 1 October 2017]).

Page 51. Scanning electron microscope images of fossil pollen grains generously provided by Dr Antoine Bercovici, (http://www.researchmagazine.lu.se/ 2014/07/22/pollen-part-of-natures-historical-archive/ [accessed 2 October 2017]). Grass phytoliths. Illustration by M. Madella (2009), from the University of Sheffield Integrated Archaeobotanical Research project (http://archaeobotany.dept.shef.ac.uk/wiki/index.php/Phytoliths_-_Interpretation [accessed 2 October 2017]), public domain under Creative Commons Noncommercial-ShareAlike 3.0 Unported (CC BY-NC-SA 3.0). Picture of oreodont teeth provided courtesy of UC Museum of Paleontology (www.ucmp.berkeley.edu). The image of Africa burning is from NASA (http://earthobservatory.nasa.gov/ IOTD/view.php?id=5800 [accessed 2 October 2017]; public domain).

Page 54. Figure compiled by the author from various sources.

Page 55. Harvester near Colusa (California) soothes his chaff-irritated skin, photograph by Joe Munroe, used by permission of the copyright-owner. From a photographic essay in Lavender D. 1976. *California – a Bicentennial History*. New York, Nashville: Norton, AASLH. It took some detective work to track down the copyright owner, the Ohio Historical Connection. Thanks for help along the way are due to Bob Beatty of the American Association for State and Local History, and to Bonnie Burgess, Joe Munroe's daughter.

Page 64. Scanning electron micrographs of rice starch from http://download.cassandralab.com/ricequality/ricequality_help/UNIMI. Cassandra.RiceQuality.html?Starchviscosityprofile.html [accessed 2 October 2017] (public domain). Diagrams of starch granules by author.

Page 65. Amino acid sequences of amylases aligned using MUSCLE Multiple Sequence Alignment tool. Molecular model of amylase produced using FirstGlance in Jmol.

Page 71. Grazing cow copyright Oleksii Terpugov (www.123rf.com/ profile_alterphoto). Cellulose model by Ben Mills (public domain: http://tinyurl.com/yafswecx [accessed 1 October 2017]).

Page 78. Kilpeck Church Green Man, from the official website of the Church of St Mary and St David (http://kilpeckchurch.org.uk/ [accessed 2 October 2017]), which so far as I know is a public domain source of pictures and text. Interestingly, the early history of the church has strong Welsh connections. The first record dates from 640 CE, when Herefordshire was part of Wales. The church was consecrated by Bishop Herewald between 1067 and 1071 under the decidedly Welsh name Llan Dewi Cilpedec. There are many versions of the drawing by Pietro Ciafferi online, none of which claims reproduction rights, which de facto places it in the public domain. *Nebuchadnezzar* from Blake Archive, Tate Britain (http://www.tate.org.uk/art/artworks/blake-nebuchadnezzar-n05059 [accessed 2 October 2017]; public domain). Photograph of the Brecon Green Man 2014 by kind permission of the Green Man Festival organisers.

Page 87. Portrait of Edmund Burke by Joshua Reynolds (https://commons.wikimedia.org/wiki/File:EdmundBurke1771.jpg [accessed 2 October 2017]; public domain). Conservative Party logo, 2006 ("It's a strong image, and of course the green in it reinforces what David Cameron is trying to communicate in terms of a more environmentally-friendly party" - Tim Montgomerie, editor of Conservative Home website).

Page 91. *The Tree Ceremony*, reproduced with kind permission of the artists, Heather Ackroyd and Dan Harvey. In the course of conversations about the present book, Heather and Dan introduced me to *The White Goddess* by Robert Graves, a work that inspires their approach to art. It's a dense but endlessly rewarding read, even though Graves begins by issuing a warning to the likes of me: 'this remains a difficult book, as well as a very queer one, to be avoided by anyone with a distracted, tired or rigidly scientific mind'.

APPENDIX 1

The next page is a summary of the geological timeline covering the period from the first terrestrial ecosystems to the present day, showing some of the notable waypoints discussed in this book. A mnemonic known to generations of geology students may be helpful:

Cows Often Sit Down Carefully. Perhaps Their Joints Creak. Persistent Early Oiling Might Prevent Painful Rheumatism (Cambrian, Ordovician, Silurian, Devonian, Carboniferous, Permian, Triassic, Jurassic, Cretaceous, Paleocene, Eocene, Oligocene, Miocene, Pliocene, Pleistocene, and Recent).

By substituting Arthritis for Rheumatism, we can accommodate the Anthropocene.

| Era | Period | Epoch | Beginning Mya | Global environment | Plant evolution | Human and animal evolution |
|---|---|---|---|---|---|---|
| Cenozoic | Present | Anthropocene | 0 | 2016 mean global surface temperature highest since instrumental recording began (mid 19th century) | First *Triticale* (wheat x rye hybrid; 1875) | Earliest watermill (3rd century BCE) |
| | Quaternary | Holocene | 0.01 | End of last N Hemisphere glaciation period (15-10 Kya) | Germination of oldest living bristlecone pine (5 Kya); domestication and dispersal of major cereal crop species (20-5 Kya) | Extinction of mammoth (4 Kya); domestication of dogs (15 Kya); dawn of agriculture (20-10 Kya) |
| | | Pleistocene | 1.5 | | Divergence of the Hordeae and the Aveneae/Poeae tribe complex, ancestors of temperate cereals and pasture grasses (1 Mya) | Use of fire, cooking, increasing hominin brain size (1 Mya) |
| | Neogene | Pliocene | 5.3 | | Olduvai Gorge date palms (1.8 Mya) | Oldowan tool-making industry (2.6-1.7 Mya); *Australopithecus* (3 Mya); Lomekwian tools (3.3 Mya); chimpanzee-human divergence (4.5 Mya) |
| | | Miocene | 23 | | Expansion of C4 grassland species (10 Mya); SE Asian, Australian C3-dominated grasslands (10-5 Mya); First C4 grasses (18 Mya) | *Sahelanthropus* (7 Mya); Human and gorilla lineages diverge (8.1 Mya) |
| | Paleogene | Oligocene | 40 | Manosque Basin, S France (30 Mya) | Rice appears (25 Mya); S American C3 grassland and associated fauna (40-35 Mya) | Human and Old World Monkey lineages diverge (31 Mya) |
| | | Eocene | 56 | | | Oreodonts (47-7 Mya); rise of ungulates (50 Mya) |
| | | Paleocene | 65 | Mass extinction event (65 Mya); Pangaea breaks up (65 Mya); major climate changes (65 Mya) | BEP-PACMAD clade diverges (60 Mya) | Human ancestral lineage diverges from the New World Monkeys (57.5 Mya) |
| Mesozoic | Cretaceous | | 145 | Alvarez catastrophe (66 Mya); Yixian Formation (127-125 Mya) | ANA group, monocots, magnoliids, basal dicots established (100 Mya); ergot-infected grass inflorescence (100 Mya); Poaceae diverge (100-70 Mya); monocot pollen (125-100 Mya); *Archaefructus* (125 Mya) | Grass phytoliths in dinosaur coprolites (66 Mya); extinction of dinosaurs (66 Mya) |
| | Jurassic | | 199 | | First angiosperms (160-140 Mya) | |
| | Triassic | | 250 | Mass extinction events (208, 245 Mya) | Conifers widely established (200 Mya); plant mass extinction event (250 Mya) | |
| Paleozoic | Permian | | 299 | | Rise of gymnosperms, decline of giant club-mosses and horsetails (260 Mya); *Ginkgo biloba* (270 Mya); | |
| | Carboniferous (Pennsylvanian) | | 323 | | Plant mass extinction event (300 Mya); *Glossopteris* (300 Mya); complex vascular and supportive tissues (308 Mya) | Gigantic insects (300 Mya) |
| | Carboniferous (Mississippian) | | 359 | | *Archaeopteris* (355 Mya) | |
| | Devonian | | 416 | Mass extinction event (367 Mya); Rhynie chert (400 Mya); Gondwanaland and N Laurussia join to make Pangaea supercontinent (400-300 Mya) | First pteridosperms (370 Mya); *Pseudosprochnus* (385 Mya); first seed plants (390 Mya); forests become dominant biome (400-300 Mya); root-like structures (408 Mya) | |
| | Silurian | | 444 | Mass extinction event (435 Mya) | Stomata (420 Mya) | |
| | Ordovician | | 490 | | Sheets of cuticle (450 Mya); first vascular plants (450 Mya); spores (475 Mya), | |
| | Cambrian | | 540 | [CO2] falls from 20x to 8x present-day values, [O2] increases from 4% to 21% (500-450 Mya); Cambrian explosion (542 Mya) | | |
| | Precambrian eon | | | | | |

INDEX

## A

**ABCE model:** also called quartet
model, a conceptual scheme in
which the identities of floral organs
are specified by the actions of
homeotic genes (qv); 36

*Abies* 83

abominable mystery 29, 34

**abscission:** active separation of a leaf,
fruit or other plant part (abscind is
the verb); 10, 11, 13

*Acer* 83

Achebe, Chinua 75

Ackroyd, Heather 90, 91

**actinomycete:** an order of bacteria
characterised by fungus-like
filamentous growth; 49

acorn 63, 83, 73

adaptive fitness 21

*Aesculus* 83

*Agave* 15

ageing 14, 16, 19, 20

*Aglaophyton* 8, 12

agriculture 1, 56, 66, 67, 68, 69, 70, 71,
76, 77, 79, 80, 81, 86, 90

**allelopathic:** descriptive of deterrence
measures plants take against
competitors, in which injurious
compounds are secreted, usually
from the roots; 26

almond 79

**alternation of generations:**
fundamental form of the plant life
cycle, comprising distinct sexual
haploid gametophyte (qv) and
asexual diploid sporophyte (qv)
stages; 7, 20

amber 48

*Amborella* 32, 35, 36

**amylase:** enzyme that breaks down
starch into glucose; 61, 64, 65

**amylopectin:** branched polymer of
glucose, component of starch; 60, 64

**amylose:** linear polymer of glucose,
component of starch; 60, 64

**ANA:** basal grade of the angiosperms,
named from its three constituent
lineages, Amborellales,
Nymphaeales, and
Austrobaileyales; 29, 35, 36

**angiosperm:** flowering plant; 4, 23, 24,
26, 27, 29, 31, 32, 34, 35, 36, 38, 40,
42, 46, 57

**annual plant:** plant that germinates,
matures, reproduces and dies in a
single growing season; 32, 51, 52,
73, 78

**anthocyanin:** pigment of the phenolic
family of compounds responsible
for the reds, purples or blues of
autumn leaves and many flowers;
83

**Anthropocene:** the most recent period
of geological time, characterised by

**cladoptosis:** process of self-pruning in which tree branches are actively shed; 11

**Clavicipitaceae:** fungal family including ergot and grass endophytes; 48, 49

**clonal plant:** member of a plant population consisting of the genetically identical separate or interconnected products of vegetative reproduction; 4, 16, 19, 22, 29, 32, 85

clover 80

club-moss 8, 24

coal 13, 24

cocoa 84

coelacanth 5

*Coffea* 83

coffee 61, 83, 84

collagen 29, 54

conifer 11, 24, 27, 31, 40

Conrad, Joseph 3

Conservative and Unionist party 87

cooking 61, 64, 66, 92

*Cooksonia* 8

cork 83

corm 43

corn 64, 73, 80

Corner, EJH 7, 33

*Cornus* 83

**cotyledon:** the rudimentary first leaf homologue present in the plant embryo; 39, 40

Cowper, John 81

cranberry 83

**Cretaceous:** geological period lasting from 145 to 65 Mya; 4, 7, 27, 29, 30, 32, 34, 36, 38, 40, 133

crop rotation 80

**cryptophyte:** a plant with resting buds that overwinter either underground or under water; 43, 49

*Cupressus* 83

cuticle 8

**cyanogenic glucoside:** type of defensive secondary compound in plants that releases hydrogen cyanide when digested by herbivorous animals; 48

cycad 27, 40

cypress 83

**cytostat:** apparatus that maintains a cell culture in the steady state; 21

# D

dairy 68, 71, 88, 90, 92

Dante 72, 78

**darnel:** poisonous annual grass, *Lolium temulentum*, a weed of cereals; 73, 74, 75, 93

Darwin, Charles 16, 29, 34

date palm 62

deciduous tree 10, 13, 45

Defoe, Daniel 33

deforestation 56, 76, 79, 82, 90

*Deliverance* 77

**Devonian:** geological period lasting from 416 to 359 Mya; 4, 7, 8, 12, 13, 30, 45, 133

diabetes 64

dicot  29, 35, 36, 39, 40, 41, 42, 48

diet  29, 47, 51, 53, 56, 58, 59, 60, 62, 65, 68, 88, 90

digestion  47, 61, 62, 64, 65

dinosaur  38, 43, 46, 47

*Diospyros*  84

**disposable soma**: evolutionary theory of biological ageing, based on the allocation of resources between reproductive and non-reproductive structures and functions in the organism's lifecycle; 21

dog  65

dogwood  83

domestication  29, 56, 65, 66, 67, 68, 69, 74, 79

dormancy  15, 49

**DPOR:** light-independent form of protochlorophyllide oxidoreductase, a key enzyme of chlorophyll biosynthesis; 31

**durian theory**: proposal by EJH Corner that ancestral durian-like tropical trees represent the evolutionary origin of arboreal lifeforms; 33

### E

El Perú-*Waka'*  70

embryo  19, 29, 41, 68

emmer wheat  66

Enclosure Movement: 76, 80, 81

**endophyte**: a bacterium, actinomycete (qv), mycoplasma (qv) or fungus

that lives within a plant for at least part of its lifecycle; 29, 49, 93

**endosperm:** starchy storage tissue of cereal grains; 68

Enlightenment  80, 81, 84

**ephemeral plant:** plant with a short life-cycle, capable of producing several generations per growing season; 26, 32, 33

*Epic of Gilgamesh*  76

*Epichloë*  49

epidermis  7

**epigenetics:** mode of inheritance in which the state of chromosomal DNA is transmitted from a cell to its progeny through sequential divisions. In this way the profile of gene repression and induction in the parent cell is perpetuated; 19

**ergot:** fungus, *Claviceps purpurea*, that grows parasitically on grasses and is the source of psychoactive alkaloids which cause ergotism, a convulsive and gangrenous condition in humans and other animals; 48, 49, 74, 117

**Eudicot:** clade of dicotyledonous plants, monophyletic in origin, divergent from the magnoliid dicot branch; 29, 35

**evapotranspiration:** return of water from land to atmosphere by evaporation and by water loss from foliage; 48

Evelyn, John  25, 72, 82

evergreen  10, 32, 35, 91

**geoxyle:** 'underground tree', in which the enlarged woody plant body is subterranean; 50

Gerard's *Herball* 62

**germline:** lineage of reproductive or generative cells, as distinct from cells of the soma (qv). Unlike animals, plants do not have separate germline and soma; 19

gigantism 10, 15, 47

*Ginkgo* 27, 40, 84

glaciation 6, 24, 67, 89

*Glossopteris* 11, 13

glucose 60, 61, 64, 68, 69, 93

**glucosidase:** an enzyme that cleaves the bond between a glucose residue and another chemical group; 61

**glycaemic:** descriptive of a food or event that increases the level of sugar in the blood; 56, 60, 61, 64, 66

**Gnetales:** order of gymnosperms close to, but not directly in, the evolutionary line leading to angiosperms; 35, 38

*Gnetum:* 40

Goethe, Johann Wolfgang von 14, 16

*Golden Bough, The* 1, 72

Goldschmidt, Richard 34

golf 90

Gondwana 4, 6, 13, 24

gorilla 53

*Gosslingia* 12

Gould, Stephen Jay 30, 34

**gramineous crop:** cultivated cereal, such as wheat or barley, derived from ancestral wild grass species; 2

grandmothering 59

grapefruit 21, 23

grassland 2, 29, 43, 46, 47, 48, 49, 51, 53, 58, 62, 67, 79, 93

Graves, Robert 72, 132

grazier 81

grazing animal 29, 43, 44, 46, 47, 48, 49, 51, 56, 59, 67, 68, 71, 75, 85

Green Man 77, 78

groundsel 33

**gymnosperm:** seed-bearing non-flowering plant, for example conifer, *Ginkgo* and cycad; 24, 29, 32, 35, 36, 40, 42, 57

# H

*Handful of Dust, A* 77

Hardy, Alister 57

Harvey, Dan 90, 91

**haustorial:** refers to an organ that absorbs nutrients or water from another structure. The cotyledons (qv) of certain species are haustorial; 40

**Hayflick limit:** the number of times a normal animal cell will divide when cultured under optimal conditions in vitro before it dies. The limit is broadly correlated with lifespan of the animal from which the cell type was isolated. Named for its discoverer, Leonard Hayflick. Plants do not obey the Hayflick rule; 20

hazelnut 63

Lowell, James Russell 93

**LPOR:** light-dependent form of protochlorophyllide oxidoreductase, a key enzyme of chlorophyll biosynthesis 31

**lycopsid:** member of the club-moss group of modern and extinct pteridophytes (qv); 8, 24

## M

**MADS-box:** a conserved DNA sequence which binds specific protein factors that regulate gene expression; 36

*Magnolia* 35, 84

maize 52, 53, 67, 69

mammoth 47, 67

mango 15

manioc 69

Manosque Basin 38

maple 11, 83

Marggraf Turley, Richard 72, 145

**Maslow hierarchy:** theory of psychological motivation proposed by Abraham Maslow; 70

meat 60, 62, 68, 71, 90

**megaphyll:** type of leaf with venation pattern and mode of attachment to the shoot axis typical of ferns and seed plants; 11

**meiosis:** reductive cell division, the necessary process in plant and animal sexual reproduction, during which the nucleus gives rise to four daughter nuclei, each with half the original number of chromosomes; 4

*Men in Black* 92

mercantilism 81

**meristem:** localised centre of active mitosis from which permanent tissue is derived. Apical meristems are located at the growing tips of shoots and roots; 4, 15, 17, 19, 20, 21, 23, 41, 42

**mesic environment:** a habitat with a moderate or well-balanced supply of moisture; 43

Meso-America 67, 84

**mesophyll:** the green photosynthetic tissue of leaves; 20

meteor 46

**microphyll:** relatively small type of leaf with unbranched midvein, found in club-mosses, horsetails and the earliest land plants; 7, 12

milk 68, 71

millet 67

Mitchell, Joni 89

**mitosis:** mode of somatic cell division in which the nucleus gives rise to two daughter nuclei with the same chromosome number as each other and the parent cell; 20

modelling 15, 18

modular structure 14, 15, 16, 17, 18, 19, 25

**molecular clock:** index of evolutionary time based on the rate of mutational change in the base sequence of DNA; 31

R

**ramet:** individual member of a
genetically uniform clonally
propagated population; 16, 19
rape 67
Raunkiær, Christen C 43
Ray, John 39
reactive oxygen species 19
Rebecca Riots 81
**reiterative growth:** mode of modular
development typical of many trees,
in which new axes proliferate as
hemi-parasitic clones; 15, 19, 17, 22
respiration 16, 17
resting bud 15, 43
rewilding 89
*Rhamnus* 83
**rhizoid:** thin hair-like structure that
functions as a root in mosses,
liverworts, ferns and early land
plants; 8
rhizome 8, 31, 43, 45, 58
*Rhus* 83
*Rhynia* 8
rice 43, 52, 53, 67
*Robinia* 83
Robinson Crusoe 33
*Robinsonia* 33
Romanticism 56, 81, 84, 87, 88, 89
**rubisco:** ribulose-1,5-bisphosphate
carboxylase-oxygenase, the $CO_2$-
fixing enzyme of photosynthesis.
Rubisco is located in the chloroplast
and is the most abundant plant
protein; 54

rumen 46, 47
ruminant 29, 47
Ruskin, John 87
rye 53
ryegrass 53, 73

S

Sahel 51
*Sahelanthropus* 53
salamander 32
saliva 61
**saltational hypothesis:** alternative to
the gradualist mechanism of
Darwinism, considering evolution
and speciation to occur by sudden
jumps as a result of large
mutations; 34
satellite imaging 51
savannah 46, 50, 53, 58, 59, 62
*Schedonorus* 48
Scott, Walter 82
**scutellum:** modified cotyledon (qv) of
the grass seed; 41
*Secale* 53
secondary compound 29, 48
**secondary growth:** increase in the
girth of the shoots and roots in
dicots as a result of the formation of
new cells in the vascular cambium
(qv); 12, 42
seed development 68
seed-bank 89
Seidel, Frederick 17
*Senecio* 33
senescence 9, 10, 14, 17, 20

streptophyte: member of the large
    clade that includes green algae and
    all land plants; 7
stress 23, 29, 30, 44, 45, 49, 54, 59, 60,
    62
sucrose 59, 61, 62, 64
sugarcane 52, 53
sugar 7, 68
sumac 83
sycamore 83
symmetry 36, 41

T

tannin: one of a group of bitter,
    astringent phenolic defence
    compounds; 47, 48, 83
tares 74
*Taxodium* 83
*Taxus* 24
tea 83
tectonics, plate: the theory of
    continental drift; 6, 24
*Teddy Bears' Picnic, The* 77
teeth 29, 46, 47, 51, 54, 59
telomere: region of repetitive DNA
    sequences at the end of a
    chromosome, that prevents
    chromosome deterioration and
    fusion; 20
temperate grass 52, 62
tepal: component of the outer whorl of
    flower structure in species such as
    tulip, in which there is no clear
    distinction between petals and
    sepals (qv); 37

*Thebaid* 76
Theophrastus 39
*Things Fall Apart* 75
Thomas, RS 1
Thünen ring 70
timber 76, 79, 81, 82
*Titus Andronicus* 75
Tolkien, JRR 77
tool 53, 56, 57, 58, 66
totipotency: capacity of a fully
    differentiated cell, tissue or organ to
    regenerate a whole plant; 20, 23
toxin 48
tracheid: tubular xylem cell, with a
    thickened lignified wall enclosing a
    cavity from which protoplasm has
    been eliminated, that forms part of
    a system for transporting water and
    mineral salts and providing
    structural support; 9, 12, 20
transdifferentiation: change in
    structure and function of a mature
    cell into another cell type; 20
transposon: mobile DNA element able
    to move from one site to another in
    the genome. Transposons can
    restructure the genome and release
    new genetic potential; 23
tree crop 79
tree fern 8
Tree of Life Project 25
tree ring 22
treehugger 90
tree-line 67
Triassic: geological period spanning
    250 to 199 Mya; 27, 30, 133

*Trichopherophyton* 12
**trimery:** of flowers, made of parts in
multiples of 3; 39
*Triticum* 53
Trojan Wars 77
*Tsuga* 84
tuber 43, 45, 58
tundra 67

## U

underground storage organ 45, 56, 58
underground tree 50
undiscovered tribe 69
ungulate 29, 47, 48
*Urpflanze* 14
**USO:** underground storage organ; 56,
58, 59, 61, 63, 66, 67, 110
**UV:** ultra-violet; 21

## V

**vacuole:** large aqueous compartment
in the plant cell, enclosed in a
membrane; 62
**vascular plant:** member of the large
grouping of land plants
characterised by the presence of a
vascular system; 7, 8, 10, 30
**vascular tissue:** system of conducting
tissue that transports water and
nutrients. Comprises xylem (qv),
phloem and cambium (qv); 8, 9, 17,
41, 42
veganism 88

**vessel:** type of xylem (qv) element,
consisting of a long chain of dead
cells with thickened walls,
characteristic of angiosperms (qv);
36
vetch 67
Vico, Giambattista 72
Vietnam War 77
Virgil 72, 73

## W

*Waldgänger* 88
walnut 11, 79, 83
Warren, John 52, 67, 84
weed 56, 73, 74, 93
Weismann, August 19, 21
*Welwitschia* 35, 40
wheat 53, 62, 73, 74
*Whispering grass* 28
*White Goddess, The* 72, 132
Whitman, Walt 52
wildfire 11, 22
willow 11
wodewose 77
*Wonderful Adventures of Nils, The* 55
wood 4, 9, 10, 12, 20, 36, 69, 72, 76, 77,
83
wool 1, 71, 80, 83
Wyclif bible 74

## X

**xylem:** tissue largely made up of
lignified dead cells that transport

water and minerals and provide
mechanical support. Wood is
mostly made of xylem cells; 9, 12,
20, 45

**xylopodia:** woody subterranean root
systems, characteristic of geoxyles
(qv); 50

### Y

Yellow and Yangtze River basins  67
yew  22
Yixian Formation  38

### Z

*Zinnia*  20
**zosterophyllous:** descriptive of a
group of extinct club-moss-like
plants of the Devonian period; 12

# ABOUT

*The War Between Trees and Grasses* is in part the continuation of themes explored in *Food and the Literary Imagination* (2014), a book I co-authored with Jayne Archer and Richard Marggraf Turley. It also includes some unfinished business left over from my most recent book, *Senescence* (2016). From the resulting brew emerged questions about the co-evolutionary relationships of people, forests and grasses, my knowledge of which was decidedly patchy. Someone (it's often attributed to Benjamin Disraeli) said 'The best way to become acquainted with a subject is to write a book about it'. So that's what I've done, and it has turned out to be an eye-opener, or maybe a mind-expander (possibly a symptom of too much starch).

I'm indebted to the many people who have, knowingly and unknowingly, contributed to this enlightening experience (but any errors, omissions, solecisms, misinterpretations and faux pas are, of course, my responsibility). The broad scope of the present book (and its title) developed from lectures I presented to a meeting in Kew Gardens in May 2016, at the invitation of the organiser, Neville Fay of Treeworks, to whom many thanks. Shortly after the Kew event, I attended a New Phytologist Symposium at the University of Bristol, on the subject of *Colonisation of the Terrestrial Environment*, which did wonders for my sketchy knowledge of the earliest land plants. I'm indebted to the organisers of, and participants in, that meeting for serving up a crash course in evolution. I met Susanna Lydon (Nottingham University) at the Bristol symposium and she kindly agreed to look at a draft of the present text. I greatly appreciate her constructively critical comments. I'm grateful to Luíseach Nic Eoin for drawing a rich vein of information on geophytes to my attention, to Sarah Lennon of New Phytologist for help with permissions to reproduce illustrations, and to Debbie Maizels (www.zoobotanica.com) for advice on artwork. Ben Thomas provided technological backup and generally kept me on my toes. Special thanks are due, as ever, to Helen Ougham who applied her editorial touch to the work as it progressed, at the same time as piloting the home life-support system. The colleagues and organisations who generously granted permission to reproduce illustrations are gratefully acknowledged in the Sources section. I am deeply appreciative of the continued support I receive from the New Phytologist Trust and Aberystwyth University.

As these final words are written, the centenary of the Welsh Plant Breeding Station, where I spent most of my working life, is less than two years away. The founding director was Sir George Stapledon (1882-1960), a great and complex man who, I think, would share many of the preoccupations that fill these pages. He was obsessed with understanding the relationship between the totality of Nature and the human urge to intervene and control. It seems fitting to end this book with Stapledon's own words on the human experience of shaping, and being shaped by, the natural world. 'Man who marches forth with fire-stick, axe or sickle', he wrote '...Man who in a few years and over wide territories can set in motion retrogressive changes of a character and of a cumulative magnitude, the likes of which ruthless Nature would never permit, or only reluctantly after the lapse of countless centuries.' (Waller 1962).

Howard Thomas,
Aberystwyth and Wye
October 2017

Howard Thomas was born and educated in Wales and, after a career in scientific research including visiting professorships at Universities in Japan, the United States and Switzerland, he is now emeritus Professor of Biology at Aberystwyth University. He has published extensively on the genetics and physiology of plant development and has a special interest in the science-humanities connection. He is a Fellow of the Learned Society of Wales and a co-author of *The Molecular Life of Plants* (2013, Wiley) and *Food and the Literary Imagination* (2014, Palgrave). His most recent book is *Senescence* (2016), a personal account of the terminal events in the lives of plants and other organisms. He is also a devout jazz musician, and is author of *20 Steps to Jazz Keyboard Harmony* (2015, Smashwords).

www.sidthomas.net/wp